# EINSTEIN'S GREATEST MISTAKE

# EINSTEIN'S GREATEST MISTAKE

▼

## Abandonment of the Aether

## (1 Dec 2005 version)

### Sid Deutsch

iUniverse, Inc.

New York  Lincoln  Shanghai

Einstein's Greatest Mistake
Abandonment of the Aether

iUniverse books may be ordered through booksellers or by contacting:

iUniverse
2021 Pine Lake Road, Suite 100
Lincoln, NE 68512
www.iuniverse.com
1-800-Authors (1-800-288-4677)

ISBN-13: 978-0-595-37481-6 (pbk)
ISBN-13: 978-0-595-67510-4 (cloth)
ISBN-13: 978-0-595-81874-7 (ebk)
ISBN-10: 0-595-37481-6 (pbk)
ISBN-10: 0-595-67510-7 (cloth)
ISBN-10: 0-595-81874-9 (ebk)

Printed in the United States of America

# CONTENTS

# PREFACE

After I "retired" in 1983, I became a Visiting Professor of Electrical Engineering at the University of South Florida, in Tampa. This gave me some time to contemplate such controversial subjects as the weirdness of quantum mechanics.

If we literally fire an electron (which we regard as a particle) through a narrow slit, it moves sideways, in general, and lands on the screen off-center. We say that it behaves like a wave when it veers off-center, and then it reverts to being a particle when it hits the screen. Similarly, if we fire a photon (which we regard as a wave) through a narrow slit, it also moves laterally, and lands on the screen off-center, as if it became a particle.

It seems to me that there is a missing ingredient, the *aether*, that is involved in the particle-wave duality of an electron and in the wave-particle duality of a photon. Streamlines in the aether guide the lateral motion of the electron or photon.

Much weirder behavior is displayed in "entanglement," when two photons are able to *instantaneously* influence each other even if they are a vast distance apart. There can be no possible explanation for long-distance entanglement, yet physicists, mature men and women, accept this as somehow realistic. In the last chapter of this book, a perfectly palatable explanation is given, based on small perturbations that accompany the aether.

Proof positive that there is an aether comes from Albert Einstein's special relativity. Imagine a planet (*THEM*) that is receding from Earth (*US*) at a speed of 100 million m/s. According to special relativity, light waves on *US* and *THEM* have a speed of 300 million m/s *relative to each planet*. If an imaginary beam of light directed at *THEM* leaves *US* at a velocity of 300 million m/s, it has to gradually speed up to 400 million m/s *relative to US* in order to land on *THEM* at a speed of 300 million m/s *relative to THEM*. Only an aether carrier can increase the velocity of light in this manner.

The first half of the book is concerned with these cosmological matters; the second half deals with the behavior of individual electrons and photons, where weird experimental results can sometimes be explained by the aether.

So what miserable fortune befell the aether? Edmund T. Whittaker had the following to say in his preface to *A History of the Theories of Aether and Electricity: The Classical Theories* [1]:

> As everyone knows, the aether played a great part in the physics of the nineteenth century; but in the first decade of the twentieth, chiefly as a result of the failure of attempts to observe the earth's motion relative to the aether, and the acceptance of the principle that such attempts must always fail, the word 'aether' fell out of favour, and it became customary to refer to the interplanetary spaces as 'vacuous'; the vacuum being conceived as mere emptiness, having no properties except that of propagating electromagnetic waves.

During the past several years, I have submitted papers based on the material in this book to various (mostly physics-related) publications. The papers were invariably rejected. Never mind the polite reasons given, I knew, from the start, that there were three strikes against me.

1.  The "big shots," following Einstein's lead, buried the aether many years ago. Its resuscitation was, simply, a preposterous proposal.

2.  I am an electrical engineer; the aether could only be revived by a physicist.

3.  A new conjecture should first be presented at conferences to get exposure to, and comments from, peer groups. I made an attempt to restore the aether at an AAAS (American Association for the Advancement of Science) convention, but it was turned down.

However, the above proof that the aether exists made the writing of this book a necessity, and, at the same time, a fun activity.

So much for the narrative about how this book came to be written. With regard to publication, I submitted a query letter to various conventional publishers, but I am afraid the subject matter was too controversial for them. Therefore, the manuscript ended up with a very unconventional publisher: iUniverse.

Their editorial evaluation group did an excellent job of making many commendable suggestions.

Many people, who went out of their way to help me, have earned my gratitude. Dudley R. Kay, head of SciTech Publishing, gave invaluable advice. Three people read the first version of the manuscript and encouraged publication: they are Dr. Raymond L. Pickholtz, George Washington University; Dr. Julio E. Rubio, Leeds University; and Dr. Tore Wessel-Berg, Norwegian University of Science and Technology. My daughter Alice, President of Bioscreen, made many helpful comments as an "intelligent layperson." My wife, Ruth, provided an environment that was conducive to creative conjecture. Finally, some of the people on my Web-site mailing list acted as informants, keeping me up to date via e-mail regarding items relating to the aether.

Sid Deutsch
Sarasota, Florida

# FOREWORD

(This section was contributed by Raymond L. Pickholtz.)

In the nineteenth century, James Maxwell formulated a set of vector equations based on the experimental work of Coulomb, Faraday, Ampere, and others who unified electric and magnetic effects. They demonstrated, among other things, that electromagnetic waves exist and travel at a finite velocity in free space. It was subsequently accepted that light was simply a kind of electromagnetic wave. These physicists had extensive experience studying waves such as sound waves, surface water waves, and waves along tubes. In each case, the very idea of a propagating wave depended on having a medium in which the waves formed and interacted. Thus, it was perplexing to Maxwell and others that light, or any electromagnetic waves, should not have a medium in space that would support them. This conjectured medium was named the aether. Maxwell labored to endow the aether with properties that would be consistent with the actual measured effects, not least of which was the finite, but extremely large, velocity of light. From the beginnings of science, the intuitive notions carried over from our daily human experiences have imposed a constraint on how we describe physical phenomenon. After Newton's great success in quantifying mechanics, and until the twentieth century, physicists viewed all physical phenomena as manifestations of mechanical systems. So it was no surprise that Maxwell and others attempted to formulate a mechanical model of the aether that permeated all space. He assumed that the aether was endowed with properties, not unlike a solid, that could be put under stress and would experience strain. He actually produced elegant work showing how the stress-strain waves in the medium called the aether could carry his electromagnetic waves through space.

The proposed all-permeating aether naturally raised an interesting possibility: the idea that we could measure the absolute velocity of the Earth as it hurtled though space (and the aether) by measuring the "aether wind" generated by the Earth's movement. Michaelson and Morley designed an ingenious experiment to measure this value using the extremely sensitive light-interferometers that Michaelson himself had pioneered. Without going into detail here, the important point is that the experiment seemed to establish that there is no aether wind in any direction of any size that could be measured.

Although Einstein was aware of the negative result generated by Michaelson and Morley, their result does not seem to have motivated his remarkable paper in 1905 that proposed the special theory of relativity. (There were a total of four papers by Einstein that year, each of which had an enormous effect on physics. His general theory of relativity, dealing with acceleration and gravity, would come later.) Einstein was driven to describe the world with elegance and simplicity. The keystone of his special theory of relativity was that light propagates with a fixed and finite velocity that is independent of relative motion. This, in itself, gave Maxwell's equations a special status in that the equations were thus held to be *invariant* to such motion. However, Einstein sacrificed Newton's mechanical theories and this led to apparently strange effects (to those who were used to basing their understanding of the universe on their everyday experiences of classical mechanics) at speeds approaching the speed of light. The effects of special relativity are now so commonplace in modern physics and even in engineering that the theory has become universally accepted.

So what about the aether? Well, Einstein and others who followed simply declared that there is no need for it, that light can travel through free space as a "field" that requires no support and, hence, no aether.

# A Short History of the Aether

**1825**: Scientists generally concluded that light is a wave rather than a corpuscular phenomenon. Since sound requires a wave carrier (a solid, liquid, or gas), they proposed that light also needs a wave carrier: the aether.

**1835**: Michael Faraday (1791–1867) introduced a model of the aether wherein it consisted of positively and negatively charged particles bound to each other.

**1842**: Christian J. Doppler (1803–1853) discovered the Doppler effect. This strengthened the view that the speed of sound does not depend on the source or its frequency, but only upon the local medium conveying the sound. Similarly, it was proposed, the velocity of light only depends on the local aether carrying the light.

**1864**: James C. Maxwell (1831–1879) derived the mathematical basis for the propagation of electromagnetic fields (EMFs). He and his colleagues assumed that EMFs have to be carried by an aether medium. The velocity of light was determined from the permeability and permittivity of the medium.

**1887**: Albert A. Michelson (1852–1931) and Edward W. Morley (1838–1923) used optical interferometry to detect the drift of the aether through their laboratory in Cleveland, Ohio. Their measurements indicated no substantial movement. This result was widely interpreted to mean that the aether does not exist, but the possibility remained that the aether is moving with the Earth.

**1887**: Heinrich R. Hertz (1857–1894) showed the propagation of radio waves, thus confirming Maxwell's equations.

**1893**: George F. Fitzgerald (1851–1901) hypothesized that a moving body becomes foreshortened along the direction of motion. This was known as the "length contraction."

**1900**: Max Planck (1858–1947) formulated quantum theory, which later became quantum mechanics. In this theory, photons are massless, but they have momentum in accordance with a "wave-particle duality" principle.

**1901**: Henri Poincaré (1854–1912) proposed various characteristics of motion through the aether.

**1904**: Hendrik A. Lorentz (1853–1928) improved on the Fitzgerald "length contraction" conjecture and proposed a "time dilation" effect.

**1905**: Albert Einstein (1879–1955) formulated the theory of special relativity and rejected the aether. He proposed that photons could propagate through the vacuum of "empty space" without an aether carrier.

**1911**: Ernest Rutherford (1871–1937) discovered the atomic nucleus. According to his theory, "solid matter" consists of nuclei that are very far apart. This leaves plenty of room for aether particles to drift through a material object.

**1915**: Albert Einstein formulated the theory of general relativity, which deals with gravitational effects.

**1919**: The theory of general relativity was confirmed by the observation of the bending of starlight as it passes near the sun.

**1924**: Louis V. de Broglie (1892–1987) proposed the "particle-wave duality" principle whereby every mass is associated with a wave.

**1929**: Edwin P. Hubble (1889–1953) revealed that the universe is expanding.

**1933**: Dayton Miller (1866–1941), and others, repeated the Michelson-Morley experiments using more accurate equipment. The conclusions are controversial. There are indications of an aether drift, but this may be caused by "noise" because the effect is relatively small. Miller reported a drift of some 0.2 million m/s (as compared with light at 300 million m/s) toward the Swordfish constellation [2].

**1964**: T. S. Jaseja et al. report on a test of special relativity using infrared masers [3].

**1979**: A. Brillet and J. L. Hall report on the use of lasers to test the isotropy of space [4].

**1983**: R. W. P. Drever et al. report on a phase and frequency stabilization scheme using an optical resonator [5]. This is the "standard" method used at the present time (2005) to measure the isotropy of space.

**2003**: Peter Wolf et al. report on a test of Lorentz invariance using a microwave resonator that was performed at the Observatory of Paris, France [6].

**2003**: Holger Muller et al. report on a modern Michelson-Morley experiment using cryogenic optical resonators that was performed at Humboldt University in Berlin, Germany [7].

**2004**: M. Consoli and E. Costanzo report on aether-drift experiments that were performed at the University of Catania, Italy [8].

**2005**: P. Antonini et al. report on a test of the constancy of the speed of light using rotating cryogenic optical resonators that was performed at Heinrich Heine University in Dusseldorf, Germany [9].

# Chapter 1

▼

# The Aether Concept

If a child challenges us to prove it, we can think of ten different ways to show that we are surrounded by air, but we are, of course, normally unaware that we live at the bottom of an "ocean" of air. It is claimed, in the present book, that we are unaware, similarly, that we are surrounded by an atmosphere of aether. There is one major difference, however: we have not been able to detect the aether.

Nevertheless, the aether provides a solution to the following mystery: how can light, or any electromagnetic wave, travel for billions of years across the vastness of the universe without losing any energy? The answer is that the universe is filled with a light-transmitting medium, the aether. The proof that there is an aether is the subject of the present chapter.

The aether concept really took off in 1864 when James C. Maxwell published his equations. They showed that electric and magnetic fields are intimately related: a changing electric field generates a changing magnetic field, and a changing magnetic field generates a changing electric field. This is the recipe for an oscillation, or wave. It revealed the theoretical characteristics of an electromagnetic wave some twenty-three years before Heinrich R. Hertz was able to propagate radio waves through the "aether." Today we can only view, with disbelief, the slow pace of scientific advancement. But remember that some 150 years elapsed between Isaac Newton (1642–1727) and Maxwell.

The symbol $c$ stands for the velocity of light. The velocity of light is 299.79 million meters/second in a vacuum. For convenience, however, the value 300 million m/s is usually used in this book.

The existence of an aether is based on the transmission of light between celestial bodies that are very far apart and receding from each other. To reduce the argument to its barest essentials, uncluttered by needless distractions, the "universe" is represented by two planets—US and THEM—in Fig. 1-1. On the left we have the Earth, or US, shown as a stationary sphere. (Please ignore, for the moment, the "aether atmospheres" surrounding the planets.) On the right, we have planet THEM, which, for convenience in drawing, is the same size as the Earth. But planet THEM is receding from the Earth at a tremendous speed.

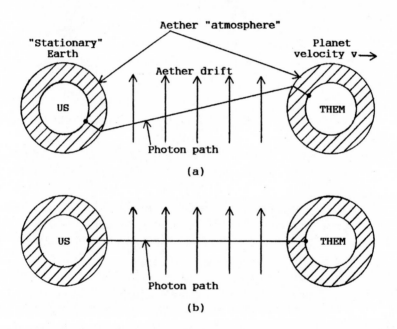

Fig. 1-1. Two possible interpretations of light-ray behavior traveling through the aether. Because of gravitational attraction, the Earth (US) and a very rapidly receding planet (THEM) each have aether "atmospheres." Interplanetary space is filled with an aether "background," shown as drifting to the north, say, at some unspecified velocity. Despite the motion discontinuity between the aether "atmosphere" and the aether drift, the density (aether particles per cubic meter) remains unchanged. (a) The photon path bends when it traverses aether motion discontinuities when leaving US, and also when arriving at THEM. (b) The photon path does not bend.

In 1929, Edwin P. Hubble discovered that the universe is expanding. In accordance with the expansion of the universe, there are many planets, stars, and galaxies that are receding from the Earth at velocities that are appreciable compared to the speed of light; in fact, objects in the outermost reaches of the universe are receding from us at *greater* than the speed of light [10], *but their local velocity of light is c.*

There is no way to prove or disprove this conjecture by communicating with a physicist on *THEM*, of course, but it does make sense. The spectral lines of each star are similar to each other and to that of our sun. [If the star is receding from *US*, the spectral lines that we receive are shifted to a lower frequency (red shift) in accordance with the Doppler equation; if the star is approaching, the lines that we receive are shifted to a higher frequency (blue shift).] The fact that the entire universe operates with the same spectral signature implies that identical natural constants, such as the speed of light, accompany each star and planet *locally*; that is, relative to each star and planet, the speed of light is 300 million m/s.

The main assertion of Albert Einstein's theory of special relativity is that the velocity of light is the same on Earth (*US*) and on the receding planet (*THEM*) *relative to each planet* [11].

Returning to Fig. 1-1, as a numerical illustration, it will be convenient to assume that planet *THEM* is receding from *US* at one-third the speed of light, or 100 million m/s.

What determines the speed of light? The characteristics of a "vacuum," the permeability and permittivity of "empty space," determine the speed of light. Apparently, the permeability and permittivity are the same throughout the universe except, possibly, for small local variations. According to Einstein, then, the speed of light in a laboratory on Earth (*US*) is 300 million m/s and, in a laboratory on planet *THEM*, it is also 300 million m/s. Obviously, this is an unbelievable situation: how can the vacuum on planet *THEM* "know" that the lab is receding from *US* and adjust so that the velocity of light is 300 million m/s even though the *THEM* lab is speeding away at 100 million m/s?

There is only one answer that makes sense: "empty space" is populated by aether particles that control the velocity of light. Furthermore, they are gravitationally bound to each planet in exactly the same way that the Earth's atmosphere is gravitationally bound to the Earth.

This answer is so unbelievable, and difficult to swallow, that the physics community insists that the correct answer is "somehow." I maintain that it is unbelievable that light (or an EMF in general) can travel from *US* to *THEM* across "empty space" without a carrier medium.

For about 100 years astrophysicists have agreed, however, that the velocity of light on each planet in Fig. 1-1 is 300 million m/s.

In the spirit of keeping things as simple as possible, the aether "atmospheres" surrounding each planet in Fig. 1-1 are portrayed as *uniform* "atmospheres," but this is undoubtedly a gross exaggeration.

It was originally believed that the aether was drifting with respect to the solar system, since it was highly unlikely that the sun was lucky enough to be the "center of the universe." In Fig. 1-1, the aether is drifting in a northward direction. Since the Earth rotates daily on its axis and annually around the sun, the aether drift should show up as the speed of light increases and decreases in synch with the Earth's motion. On the other hand, if the aether is gravitationally bound to the Earth, like an atmosphere, the aether drift as measured in an Earth laboratory should be zero. This not so, however, if we travel in any direction, especially if we leave the Earth in a space vehicle: when the vehicle exits the putative aether atmosphere, we should be able to measure the true aether drift as it is modified by the vehicle's motion with respect to the Earth or sun.

Despite all of this motion and commotion, the aether *density* remains constant; that is, the aether particles are evenly distributed. How can this be possible? Think of a glass of water in which the upper half is gently stirred (to duplicate the slow aether drift relative to the speed of light). At the boundary between the upper and lower halves of the glass, the molecules of water rush past each other, but there is a negligible change in density (number of molecules per cubic meter). Admittedly, this peaceful scenario would change with violent stirring, but aether drifts are slow relative to the speed of light. As pointed out earlier, in A Short History of the Aether, Dayton Miller measured an aether drift of less than 0.1% of the speed of light [2].

What we need is a small, lightweight piece of equipment that can be carried by a space vehicle. Today, with our highly accurate and sophisticated technology for measuring frequencies, it should be possible, although very difficult, to measure the speed of light *accurately*, to within a few meters per second. Until this is done, the aether will remain an elusive concept.

Whenever the experiments aboard a space vehicle are listed, I look for an instrument that can accurately measure the velocity of light. I believe that there is no such instrument, so it would make a great PhD thesis topic for an engineering student.

But we do have very accurate equipment—interferometers—for comparing the speed of light from south to north, say, to the speed of light from east to west. This should be capable of detecting the aether drift due to a horizontal motion of, say, 1 meter/second. This is further discussed in Chapter 6.

## Traveling with a Photon

Let's try a different perspective. In Fig. 1-1(b), let's travel with a photon as it traverses its "photon path" (while planet *THEM*, remember, is receding at 100 million m/s). The photon leaves *US*, as part of an idealized laser beam, flying to the right at 300 million m/s. When it gets beyond the aether "atmosphere," into the "aether drift" area, it continues to fly at approximately 300 million m/s. But when it gets to the *THEM* planet's aether "atmosphere," which is moving to the right at 400 million m/s, the photon speeds up from 300 to 400 million m/s relative to *US*. It lands on *THEM* planet at a velocity of 300 million m/s relative to *THEM*. Without an aether, the photon would land at a speed of 200 million m/s relative to *THEM*, which is of course impossible because it violates the universal electromagnetic wave speed of $c = 300$ million m/s.

When the photon speeds up from 300 to 400 million m/s, it is a wave that becomes stretched, so the frequency decreases (the Doppler effect) by a factor of 3/4. In other words, an originally blue photon $(f = 6 \times 10^{14}$ Hz$)$ would actually appear as a red color, in this case $f = 4.5 \times 10^{14}$ Hz, when it is received. The conventional Doppler equation for cosmological objects is

$$f_{\text{generated}}/f_{\text{received}} = 1 + (v_{\text{receding}}/c) \quad (1\text{-}1)$$

so that, in the present case, $f_{\text{generated}}/f_{\text{received}} = 1.333$. A cosmologist would say that this corresponds to a red-shift $z$ of 0.3333.

Relative to the people on *THEM*, the Earth (*US*) is receding at 100 million m/s to the left.

In the usual scenario, *US* astronomers would receive an originally blue photon from *THEM*. Because of the expansion of the universe, however, it reaches *US* at a lower frequency (i.e., the red shift has occurred).

If the *THEM* planet approaches *US* at a speed of 100 million m/s, a blue photon leaving *US* would slow down to 200 million m/s, relative to *US*, when it reaches *THEM*. The photon becomes compressed as its frequency increases by a factor of 3/2. In this event, an originally blue photon would actually change to ultraviolet, with $f = 9 \times 10^{14}$ Hz.

In the original Doppler effect, describing the transmission of sound, the source moves, but the local air is stationary with respect to the observer. The cosmological Doppler effect, describing the transmission of light, is fundamentally different because the source moves but the intervening aether *also* moves. Therefore, if $v_{\text{receding}}$ (or $v_{\text{approaching}}$) exceeds 100 million m/s, a relativistic Doppler effect equation should be used in place of Equation 1-1 (see page 203 of [12]).

Photons ignore each other, and two photons that hit each other head-on only yield the algebraic sum of their respective wave packets. Following the "collision," they continue to propagate, unchanged, at the speed of light. The wave-particle duality of quantum mechanics does not show up here; colliding photons certainly do not act like particles. Since they behave like waves, which require a carrier, this can be taken as further evidence for an aether carrier medium.

A bit of mystery remains, however. The energy of a photon (in joules) is given by

$$E = hf, \quad (1\text{-}2)$$

where $h$ is Planck's constant ($6.6261 \times 10^{-34}$ joule-seconds) and $f$ is the frequency (Hz) of the photon. Since the energy of the photon is proportional to the frequency of the photon, the photon gives up some energy when its speed goes from 300 to 400 million m/s, relative to *US*, because its frequency decreases. What happens to the lost energy? The answer is to be found in a ball thrown upward: its kinetic energy is converted into potential energy, not into heat. When the ball falls back down, its potential energy is returned as kinetic energy. Analogously, the energy lost by a "blue" photon is converted into "potential energy" carried by the "red" photon. How can the red photon convert its potential energy back into a blue photon? Obviously, if an object *approaches* the Earth, and you return the red photon (via a mirror), it will reach the approaching object as a blue photon. (Approaching and receding velocities have to be the same, of course, for an exact exchange of energy.)

In 1905 the red shift (or blue shift) was unknown, of course, so Einstein didn't have to deal with it (until 1929, when Hubble revealed the expansion of the universe).

Hopefully, photon energy is somehow converted into protons in black holes, thus rejuvenating matter in the universe.

Although it is not a proof, it is conceptually palatable to say that the expansion of the universe is the expansion of space that is occupied by the aether. Beyond the aether "cloud," there is an undefined lifeless nothingness, undisturbed by electromagnetic waves that cannot propagate. This perspective is illustrated in the vastly oversimplified model of the universe in Fig. 1-2(a). The universe is 14 billion years old (according to an authoritative article in a 2005 issue of *Scientific American*). The outer surface of the sphere representing the universe has been expanding at the velocity of light, $3 \times 10^8$ m/s, for 14 billion years; this can be interpreted as an aether "drift." During the expansion, all of the cosmological objects in the universe have been moving away from each other. Is the expansion noticeable during a human lifetime?

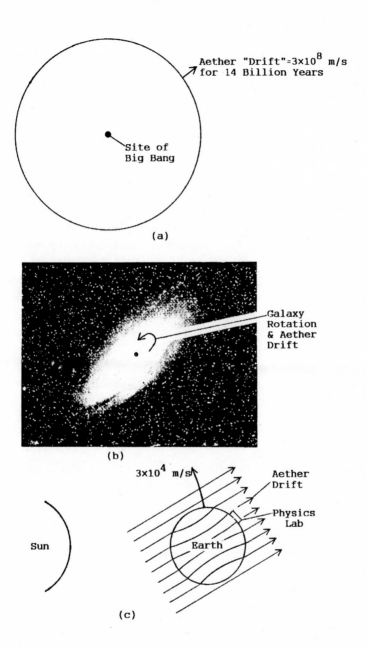

**Fig. 1-2.** Conjectures concerning the aether drift shown in various contexts. (a) As the medium that defines the spherical universe, expanding at the velocity of light for 14 billion years. This corresponds to the expansion at 7 angstroms/second of a rod that is $3 \times 10^8$ meters long. (b) In a galaxy. (c) Streamlines near and through the Earth.

Suppose we imagine a rod that is $3 \times 10^8$ meters long (the distance light travels in one second). How much does it expand in one second? We have to first find the age of the universe in seconds. According to Table A-2 (found in the Appendix), 1 year = $3.1558 \times 10^7$ seconds. Multiplying by 14 billion years, we find that

$$(3.1558 \times 10^7) \times (14 \times 10^9) = 4.418 \times 10^{17} \text{ seconds}$$

is the age of the universe. Then expansion of the rod is given by ($3 \times 10^8$ meters)/($4.418 \times 10^{17}$ seconds) = $7 \times 10^{-10}$ m/s, or 7 angstroms/s. This brings the expansion of the universe "down-to-earth" since we can visualize a rod $3 \times 10^8$ meters (or 186,000 miles) long expanding 7 angstroms (about the diameter of two water molecules) per second. One can conclude that the expansion of the universe is accompanied by minuscule changes in permeability and permittivity.

Aether drift in a galaxy is portrayed in Fig. 1-2(b). Looking down, the galaxy is rotating counterclockwise. In this book, to satisfy the requirements of special relativity, the aether is represented as being gravitationally bound to massive structures. Therefore, the aether is drifting counterclockwise along with the stars of the galaxy. (This is also reminiscent of the dark matter in the galaxy.)

Figure 1-2(c) shows aether drift streamlines near and through the Earth as the Earth's orbit carries it around the sun at a speed of $3 \times 10^4$ m/s. Aether particles are assumed to be the size of electrons or smaller, so they fill the voids between neutrons, protons, and electrons. They interact with matter, symbolically depicted by streamline curvature.

## Current Research

Is there a conflict between the aether atmosphere of Fig. 1-1 and the aether drift of Fig. 1-2(c)? Not at all. The atmosphere is stationary with respect to the Earth, and that is why it has not been detected. The aether drift is relatively slow, superimposed on the atmosphere; with sufficiently sensitive instruments, it should be possible to detect the aether drift. The titles of references [2] through [9], which are discussed in Chapter 6, chronologically tell the story of increasing instrument accuracy:

[2] "The Aether-Drift Experiment and the Determination of the Absolute Motion of the Earth," 1933;

[3] "Test of Special Relativity or of the Isotropy of Space by Use of Infrared Masers," 1964;

[4] "Improved Laser Test of the Isotropy of Space," 1979;

[5] "Laser Phase and Frequency Stabilization Using an Optical Resonator," 1983;

[6] "Tests of Lorentz Invariance Using a Microwave Resonator," 2003;

[7] "Modern Michelson-Morley Experiment Using Cryogenic Optical Resonators," 2003;

[8] "From Classical to Modern Aether-Drift Experiments: The Narrow Window for a Preferred Frame," 2004;

[9] "Test of Constancy of Speed of Light with Rotating Cryogenic Optical Resonators," 2005.

What have the instruments uncovered thus far? Unfortunately, all of them are stationary (relying on the Earth's rotation with respect to a physics laboratory) or slowly rotating; they cannot detect the stationary aether atmosphere. Aether drift has not been reliably unearthed. However, an instrument that swings back and forth, like the weight of a pendulum, creates its own aether drift, in effect, and should yield meaningful non-zero readings. By far the greatest experiment would be a trip in a space vehicle.

Marcus Chown has summarized some of this research in his article, "Catching the Cosmic Wind" [41]. The first four paragraphs of his article have been reproduced here:

HEADLINE: Catching the cosmic wind; We killed off the aether a hundred years ago. So why is the search back on, asks Marcus Chown?

Two hundred thousand dollars seems a small price to pay. If the most famous null result in science was right, at least we'll finally be sure. And if it was wrong, then Einstein is no longer king of the universe. No wonder Maurizio Consoli is keen to get started. This experiment could be dynamite.

Consoli, of the Italian National Institute of Nuclear Physics in Catania, Sicily, has found a loophole in the 19th-century experiment that defined our modern view of the universe. The experiment established that light always travels through space at the same speed, whatever direction it is heading in and whatever the motion of its source: there is no way to put the wind in light's sails.

Einstein used this foundation to build his special theory of relativity, but it seems his confidence may have been premature. Consoli's paper, published in Physics Letters A (vol 333, p 355), shows that there might be a wind that blows in light's sails after all: something called the aether.

Until just over a century ago, most physicists believed that this ghostly substance filled all of space. Their reasoning was straightforward enough: the prevailing opinion was that light traveled as a wave, just like sound. And just like sound waves, light waves would need something to move through. Light, they believed, was the result of oscillations in the aether.

Since this chapter is devoted to the "aether concept," it is appropriate to recite other viewpoints related to the aether concept. In particular, an extension to gravitational effects is advocated by Maurice Allais [13]. (See his Web site: http://allais.maurice.free.fr/English/aether1.htm.) His report, "About the Aether Concept," is repeated below, minus some mathematical sections and references. What Professor Allais has to say is interesting and important and, besides, I agree with much of it. An introduction states the following: "[T]his site has been created by several of his pupils and admirers, and is dedicated to him. Being at the same time physicist and economist, he succeeded in printing his mark in these two disciplines, and he is the only French economist to have obtained the Nobel Prize of Economic Sciences (1988). As well in Physics as in Economy, he passed his life to seek the unusual one, the exceptional one, and to rectify what there could be wrong in the generally accepted ideas."

About the Aether Concept (24 July 2003)
Some points seem essential concerning the concept of aether and its evolution over time.

**I.—No action at a distance.**

1. No action at a distance is conceivable without the existence of an intermediary medium.
All known actions, gravitational, optical, electromagnetic, propagate through a *medium*, the aether.
2. The attraction according to Newton's law of the inverse square of the distance or Ampere's formulas *are not actions at a distance. They result from local actions* that propagate progressively across space through the aether.

- - - - - - - - - - - - - - - - - - - - - - - - - - - - - - - - - - - - - -

**III.—Interferometric experiments of Michelson and Miller**

4. Assuming the aether as motionless, isotropic and euclidian, the interferometric experiment of Michelson should have revealed, for the speed of the earth relative to the aether, a value of about 30 km/s.
Since a speed of the order of 8 km/s was recorded, it was deduced wrongfully that the outcome was due to errors of observation and that actually it was impossible to record the earth speed on its orbit, an axiom put as the basis of the special theory of relativity.

In fact, I have shown in 1999 *quite extraordinary regularities* in Miller observations, which are totally impossible to attribute to some perverse effects like temperature.

As a result it follows, *a radical and definitive collapse* of the theory of relativity.

5. In fact, what Miller experiments show is a variation of the speed of light due to a local anisotropy of the aether, *which is quite different* from a direct relation between the earth movement and the speed of light.

There is no "*aether drift*" at all.

The Millers interpretation of his experiments is *totally erroneous*. Suffice for me to recall that this interpretation completely neglects the average deviations of the azimuth relatively to the meridian and doesn't explain the assumed reduction by Miller of the effects' amplitude.

Actually, the talk about the aether drift due to the earth movements comes from the old and obsolete concept of action at a distance. If drift there is, it is a local drift of the earth due to the aether's environment.

### IV.—Aether movements and deformations

6. Contrary to what was assumed in XIX[th] century and early XX[th] century, the aether is subject to movements and local deformations; in other words, the aether is an anisotropic medium. This anisotropy varies over time and space.

The properties of the "*vacuum*" are nothing else than aether properties.

7. The movements and deformations of the aether influence the different phenomena observed and all these phenomena are influenced in the same way.

8. Atoms, particles, photons…are but (local) singularities of the aether that remain to be explained by differential equations.

### V.—Periodical structure of the anisotropy of the aether

9. Movements and distortions of the aether have a periodic structure that we find particularly in the observations on gravity and optics.

It follows that all observable physical phenomena have the same periodical structure.

This explains accordingly the phase agreement between the paraconic pendulum movements and the optical observations on fixed sights and collimators.

10. All the anomalies found, whether those related to the paraconic pendulum movements or those of optical views on fixed sights and collimators or those of interferometric measurements, have a periodic structure. All are related to the periodic anisotropy of the aether.

Their cumulative effects cancel each other over time. As a result they do not invalidate the fundamental structure of celestial mechanics.

What these anomalies show is the existence of complementary terms that remain to be formulated and that relate to the movements and distortions of the aether.

## VI.—Differential equations of the aether movements

11. The usual representation by differential equations of the gravitational, electrostatic and electromagnetic fields hide *the fundamental identity* underlying all these phenomena.

This fundamental identity is the only physical reality. The task today is to determine the whole set of differential equations representing this reality of which the actual equations of the treatises of physics are but special cases.

12. What is missing today is a clear and comprehensive representation of physical reality; i.e., of the fundamental properties of the aether, a representation that no mathematical abstraction in nowaday's texts of physics can bring about.

13. Nature doesn't leave any room to chance and all is determined by cause and effect relationships.

What's called haphazard is nothing but a representation of our ignorance. But the permanent nature of the statistical laws shows the existence of a hidden order.

## VII.—The predominance of facts over theories

14. The history of physics is extremely instructive but couldn't suffice. The thoughts of the greatest geniuses of the past couldn't apply but over the experimental facts known at the time. Their writings are irreplaceable but they cannot be considered as sufficient.

More evidence of the existence of the aether can be found in the weird results accompanying "entanglement" and the diffraction of single photons and electrons in a double-slit apparatus. This evidence is considered in the second half of the book. The first half is devoted to cosmological considerations.

▼

# CHAPTER 2

# ALBERT EINSTEIN'S TRANSGRESSION

It is currently common knowledge that the universe is expanding. But around 1917, Albert Einstein (along with other astrophysicists) was convinced that the universe was "flat," not expanding or contracting. Accordingly, Einstein added a term to stabilize the expansion-contraction equation. In 1929, however, Edwin Hubble revealed that the universe was, in fact, expanding. Einstein's comment, with regard to the term he had added to the equation, was that this was his "greatest blunder" [14].

But this was not really a serious mistake because the expansion of the universe is an ongoing topic in cosmology; a major change was introduced as recently as 1998. (Einstein died in 1955 at the age of seventy-six.)

Here I shall argue that Einstein committed a far greater transgression—he created all of the conditions that necessitated an all-pervading aether, and then he abandoned it!

The propagation of sound requires a material medium—atoms or molecules. Without a carrier, sound cannot pass through a vacuum. Analogously, around 1864, the aether became necessary to James Clerk Maxwell and his contemporaries. It is appropriate for the purpose of this book to repeat the following aether concept: an electromagnetic field (EMF) is an ensemble of minuscule photons. Photons can propagate for billions of years, through the vastness of the universe, at a velocity of $2.998 \times 10^8$ m/s, independent of photon frequency and with zero attenuation. There is no way this is possible without an all-pervading aether carrier. The aether is most peculiar and special in that it can transport a photon, with energy equal to $hf$ (Equation 1-2), without the frictional loss that characterizes a sound wave.

At an atomic and subatomic level, friction loses its macroscopic characteristics. An electron rotates endlessly around a nucleus, an impressive feat because it is taking place in an ocean of aether particles. Our macroscopic image of aether particles moving aside, to allow the electron to proceed, and closing ranks after the electron has passed, is surely unrealistic and oversimplified. The subatomic world continues to elude us.

### Why Einstein Rejected the Aether

Some background information on the aether, as well as Albert Einstein's rejection thereof [16], is given by Peter Galison [15] (pp. 14–15):

> Einstein began his relativity paper ["On the Electrodynamics of Moving Bodies"] with the claim that there was an asymmetry in the then-current interpretation of electrodynamics, an asymmetry not present in the phenomena of nature. Almost all physicists around 1905 accepted the idea that light waves—like water waves or sound waves—must be waves *in* something. In the case of light waves (or the oscillating electric and magnetic fields that constituted light), that something was the all-pervasive *aether*. Most late-nineteenth-century physicists considered the aether to be one of the great ideas of their era, and they hoped that once properly understood, intuited, and mathematized, the aether would lead science to a unified picture of phenomena from heat and light to magnetism and electricity. Yet it was the aether that gave rise to the asymmetry that Einstein rejected.

This asymmetry is illustrated in Fig. 2-1. In (a), we have the electric field $\underline{E}$ induced by a stationary magnetic field, $\underline{B}$, along a non-conducting rod that is cutting across the $\underline{B}$ flux lines at a velocity $\underline{v}$. The simple equation is

$$\underline{E} = B v \quad (2\text{-}1)$$

using MKS units (volts/meter, webers/sq. meter, and meters/second) [17] (p. 44). A weber/sq. meter is also called a tesla. Various right-hand and left-hand rules are used to get the correct polarity.

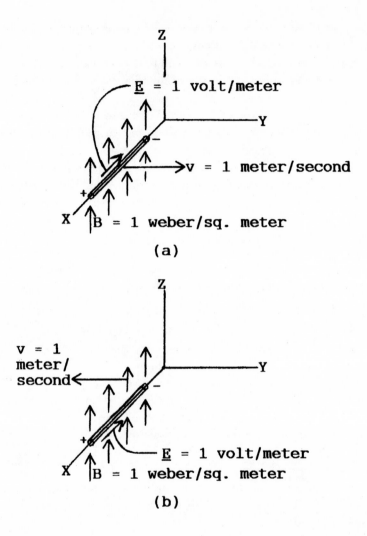

**Fig. 2-1.** The two basic ways in which an electric field, E, can be induced along a non-conducting rod. (a) The rod cuts across the B flux lines, of a stationary magnetic field, at a velocity v. (b) The magnetic flux field B, cuts across the stationary rod, at a velocity v.

In Fig. 2-1(b), paraphrasing the above, we have the electric field $\underline{E}$ induced by a moving magnetic flux field $\underline{B}$ as it cuts across a stationary non-conducting rod at a velocity $\underline{v}$. In (a), the rod moves; in the politically correct jargon of 1905, one would say that the magnetic field is stationary while the rod moves *with respect to the aether*. In (b), we would say that the rod is stationary while the magnetic field moves *with respect to the aether*. This is the asymmetry that Einstein rejected because the induced electric field is the same in (a) and (b); obviously, it is the *relative* motion that is decisive.

Electrical engineers have always used the relative motion, of course, in calculating induced voltages. But let us return to the milieu of 1905. Einstein used a coil instead of a non-conducting rod, but the principles are the same. As Galison points out on page 15, "the aether itself could not be observed, so in Einstein's view there was but a single observable phenomenon: coil and magnet approach, producing a current in the coil (as evidenced by the lighting of a lamp)" [15]. But even if the aether *could* be observed (and, with the instruments currently available, this is a possibility), it would not alter the conclusion with regard to relative motion.

From the vantage point of 2005 rather than 1905, it seems that Einstein's rejection of the existence of the aether was based on unimportant considerations because practitioners used relative motion anyway. But he applied relativity as a general philosophy, ending up with special relativity, which destroyed earlier concepts of time and space.

My own view is that every massive object (planets, moons, suns, etc.) has a stationary aether "atmosphere" that is held in place by gravitational attraction as depicted in Fig. 1-1. Therefore, there is nothing wrong with saying that the rod or wire or magnetic field moves with respect to the aether. To discover where and how the aether "atmosphere" ends, as pointed out in Chapter 1, we need a small, lightweight piece of equipment that can be carried by a space vehicle and that can measure the speed of light *accurately*, to within a few meters/second.

It seems likely that Einstein's rejection of the aether had much to do with his antiauthoritarian personality. Galison had the following to say with regard to Einstein's schooling (p. 228): "After a first unsuccessful application, Einstein began his training at Switzerland's (and one of Europe's) great technical universities: the Eidgenossische Technische Hochschule (ETH), founded in 1855. Certainly the ETH of 1896 was a very different place from the Ecole Polytechnique that Poincaré had entered in the early 1870s" [15]. As a young man of Jewish ancestry, he was unquestionably discriminated against. In 1901 (p. 233), "...authorities were not about to respond...with a shower of job offers. One after another, rejections arrived, including one for the position of senior teacher, Mechanical Technical Department in the Cantonal Technikum at

Burgdorf" [15]. But, Einstein was finally offered (p. 233) "a genuine prospect of employment. The Swiss Patent Office in Bern had placed an advertisement for an opening" [15]. Einstein worked at the Patent Office from June 1902 to October 1909 (age thirty).

With regard to Einstein's personality, Galison had the following to say (p. 46): "Framing himself as a heretic and an outsider, Einstein scrutinized the physics of the fathers not to venerate and improve, but to displace." Later, Galison asserted that (p. 232) "Einstein's relentless optimism and self-confidence, combined with a biting disregard for complacent scientific authority, shows in a myriad of letters." Galison also states that (p. 297) "[f]or the young Einstein, repair held little appeal. Tearing down the old was a bracing pleasure. While Poincaré maintained the aether as crucial in his 1909 Lille address, Einstein began a talk of his own at almost exactly the same time with a specific reference to a physicist (not Poincaré) who had assessed the aether's existence to 'border on certainty.' Then Einstein knocked the author's assertion into the trash." In conclusion, Galison asserts that (p. 310) "[d]elightedly mocking senior physicists, teachers, parents, elders, and authority figures of all kinds, happily calling himself a 'heretic,' proud of his dissenting approach to physics, Einstein shed the nineteenth century's aether with an outsider's iconoclastic pleasure."

## Conjecture about Einstein's Rejection

I think there is more to it than the arguments presented by Galison. The following argument is *entirely conjecture* on my part. In special relativity, Einstein postulated that the velocity of light on, say, the planet in Fig. 1-1 that is rapidly receding from the Earth, is the same as the velocity on Earth; that is, 300 million meters/second. He saw that the obvious physical explanation was that the receding planet, and the Earth, and every massive structure, had an aether "atmosphere" held in place by gravitational attraction. But there was no way to prove this. The Michelson-Morley experiment was not able to detect an aether. The aberration of starlight was not able to detect an aether. Einstein wisely saw that he would be bogged down in endless speculation about the aether, an elusive, ghost-like material, so he simply abandoned it. How did electromagnetic waves propagate through the vastness of the universe? Somehow, but they did not need an aether carrier!

The aether was banished from Einstein's writings from 1901 on (except to denounce it). But in a talk in 1920 titled "Aether and the Theory of Relativity," he softened his stance [18]. Here is the final paragraph (see the Web site http://www.mountainman.com.au/aether_0.html):

Recapitulating, we may say that according to the general theory of relativity space is endowed with physical qualities; in this sense, therefore, there exists an aether. According to the general theory of relativity, space without aether is unthinkable; for in such space there not only would be no propagation of light, but also no possibility of existence for standards of space and time (measuring-rods and clocks), nor therefore any space-time intervals in the physical sense. But this aether may not be thought of as endowed with the quality characteristic of ponderable media, as consisting of parts that may be tracked through time. The idea of motion may not be applied to it.

In my opinion, Einstein hems and haws about the aether, but please read the entire text of his talk and judge for yourself.

Gradually, the aether concept faded as nobody could prove that it existed.

# CHAPTER 3

▼

# MOSTLY ABOUT SOUND

It is useful to review certain characteristics of an electromagnetic field (EMF). Because some of these are descriptive of *any* wave, the discussion is reinforced if we first consider a wave that is completely different in some respects, but with which we are thoroughly familiar: that of sound.

I remember, many years ago, when my classroom teacher placed a ringing bell under a jar. He attached a small pump to the jar and proceeded to remove the air from the jar. As the volume of air in the jar decreased, the loudness diminished, showing that air in the jar was necessary to transmit the bell's sound to the student audience in the room.

What is a vacuum? The absence of air? Nothing? If the above experiment is repeated with a magnet placed across the jar, it will turn out that removing the air in the jar has no effect upon the magnetic field in the jar. Similarly, if the experiment is repeated with electrodes attached to a battery placed across the jar, then it will turn out that removing the air from the jar has a negligible effect upon the electric field in the jar.

It would certainly help if we knew what magnetic and electric fields really are, but the fact nevertheless remains: a vacuum is more than "nothing." It can sustain magnetic and electric fields. This much was known by physicists in 1864. At that time, James Clerk Maxwell presented the equations that describe an electromagnetic field. According to Maxwell, a changing magnetic field generates a changing electric field, which in turn generates a changing magnetic field, and so on and so on.

Despite their differences, however, the characteristics of sound and EMF propagation can be presented in the *same* table, Table 3-1. In the bottom half of the table, some of the characteristics of a "vacuum" and its EMF passengers are listed. The permeability and permittivity of a "vacuum" really describe the properties of the aether; it is well to remember that this is the aether that we are urged to discard.

**Table 3-1.** Analogies between sound and an electromagnetic field (EMF). Air and water values are valid at a pressure of 760 mm Hg and a temperature of 0°C. Definitions for the symbols used are listed below:

$\rho_D$ = density (kilograms/cubic meter);
$Y_0$ = modulus of elasticity (pascals);
$v$ = velocity (meters/second);
$Z_0$ = characteristic impedance (ohms);
$\mu$ = permeability (henries/meter);
$\varepsilon$ = permittivity (farads/meter).

| | Given values | | Derived values | |
|---|---|---|---|---|
| **Sound** | $\rho_D$ (kg/m³) | $Y_0$ (Pa) | $v$ (m/s) | $Z_0$ ($\Omega$) |
| Air | 1.297 | $1.425 \times 10^5$ | 331 | 430 |
| Water | 992 | $0.232 \times 10^{10}$ | 1529 | $1.517 \times 10^6$ |
| Nickel | 8700 | $20 \times 10^{10}$ | 4795 | $4.171 \times 10^7$ |
| **EMF** | $\mu$ (H/m) | $1/\varepsilon$ (m/F) | $v$ (m/s) | $Z_0$ ($\Omega$) |
| Vacuum, air | $1.257 \times 10^{-6}$ | $11.29 \times 10^{10}$ | $2.998 \times 10^8$ | 376.7 |
| Ruby mica | $1.257 \times 10^{-6}$ | $2.092 \times 10^{10}$ | $1.290 \times 10^8$ | 162.1 |
| Water | $1.257 \times 10^{-6}$ | $0.1448 \times 10^{10}$ | $0.339 \times 10^8$ | 42.66 |

A sound wave, illustrated in Fig. 3-1, is a longitudinal vibration. When a sound wave propagates, the molecules (of air, for example) vibrate back and forth in the same direction that the wave travels. An electromagnetic wave is a transverse vibration: that is, the electromagnetic field lines are oriented at right angles to the direction in which the wave travels. An EMF is transmitted without losses through a vacuum, and all EMFs travel at the speed of light, $c = 2.998 \times 10^8$ m/s in a vacuum. By contrast, a sound wave is transmitted through matter (gas, liquid, or solid); the velocity of propagation is different for every medium, and the velocity is relatively slow. However, both waves are analogous with regard to the equations for the velocity of propagation ($v$) and characteristic impedance ($Z_0$).

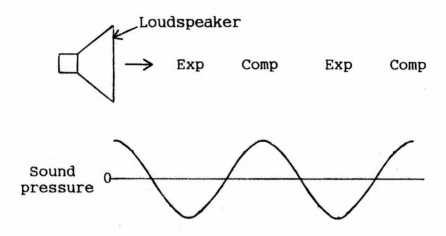

**Fig. 3-1.** Expansions and compressions of a longitudinal sound wave.

## Characteristic Impedance

What is characteristic impedance? It informs us how well a signal can be transmitted from one medium to another without suffering a loss due to reflection. For example, because of the huge difference in the $Z_0$ of air and water (430 versus 1,517,000), an underwater swimmer is shielded from sound in the air above. The sound is almost completely reflected back, as if the water surface were a mirror. For a visible-light wave striking a mirror, the situation is more complicated because the reflecting surface is an electrical conductor (silver or aluminum). Nevertheless, the idea is the same: at the sharp discontinuity between air and the silver or aluminum coating, visible-frequency EMFs ($4 \times 10^{14}$ to $7.9 \times 10^{14}$ Hz) are almost completely reflected.

For sound, the velocity $v$ is determined by $Y_0$ (Young's modulus of the medium) and $\rho_D$ (the density of the medium). Young's modulus is the stress of the medium divided by the strain on the medium; that is, given a cube of the material in which one side of the cube forms a piston, we apply a force and measure the movement of the piston. The stress has units of newtons per square meter; the strain is the change in thickness (meters) divided by the original thickness (meters), so the strain is "dimensionless" (a fraction without units). Then $Y_0$ has the units of newtons per square meter, which is often represented by the pascal, whose symbol is Pa.

One can think of Young's modulus, also called the modulus of elasticity, as a measure of the stiffness of the material. Table 3-1 lists $Y_0$ for three common and representative materials: air, water, and nickel. Young's modulus is relatively small for air, of course. We see that water, frequently cited as "incompressible," is much more compressible than a metal such as nickel.

The density of the medium affects the velocity of the sound waves because the movements associated with the compression and expansion of the medium, shown in Fig. 3-1, are opposed by the inertia of the medium. At one extreme, Table 3-1 lists air, which has a low density, a high compressibility, and a low velocity of sound (331 m/s). At the other extreme, Table 3-1 lists nickel, which has a high density, a low compressibility, and a high velocity of sound (4795 m/s). Water falls in between these two extremes.

At the right side of Table 3-1 are two columns of "derived values." It turns out that the equations for sound and EMF are similar, thus supporting the notion that the EMF is conveyed by the aether analogously to the way sound is conveyed by atoms and molecules. This is further discussed in the Appendix.

▼

# CHAPTER 4

# AN ELECTROMAGNETIC FIELD

For the EMF, only the simplest example, that of a plane wave, is considered below. Accordingly, in the end view of Fig. 4-1, the EMF consists of vertical electric ($\underline{E}$) and horizontal magnetic ($\underline{H}$) fields that are mutually perpendicular (at right angles to each other) and perpendicular to the direction of propagation. The fact that the fields are mutually perpendicular means that they form a plane wave. Why is it necessary to have such a complicated drawing for the EMF when the simple diagram of Fig. 3-1 suffices for sound? We can get away with Fig. 3-1 because sound is much simpler: it is a one-dimensional vibration that vibrates in the direction of propagation. Also, it is important to get the full flavor of an EMF before we break it down into its individual components, photons such as those shown in Fig. 4-2.

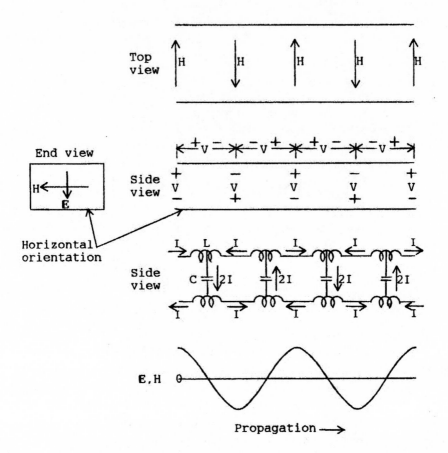

**Fig. 4-1.** "Photograph," taken at a particular instant of time, of the electromagnetic field (EMF) associated with a waveguide. Although the *L*s and *C*s formed by the walls are distributed, it is more convenient to show them as discrete elements that are located at the zero crossover points of the electric (<u>E</u>) and magnetic (<u>H</u>) fields. The <u>E</u> and <u>H</u> fields are mutually at right angles to each other and to the direction of propagation. The arrows show the conventional direction of the electric field from + to −, or the direction of current (not electron) flow.

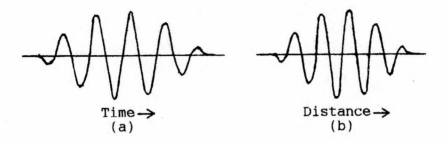

Time → 
(a)

Distance → 
(b)

**Fig. 4-2.** The wave packet representation of a photon. (a) The electric (or magnetic) field measured at a particular point in space. The photon flies by at the speed of light, $c = 2.9979 \times 10^8$ meters/second (in a vacuum). The wiggles occur at a frequency $f$. (b) A "photograph" taken at a particular instant of time. The "size" of a photon is unknown.

In order to show the $\underline{E}$ and $\underline{H}$ fields, Fig. 4-1 uses a waveguide, which is conveniently chosen to be two wavelengths long, to propagate the EMF. The waveguide is a hollow, rectangular bar made out of a good conductor (to minimize electrical losses). To the EMF signal, however, the waveguide is much more than a hollow bar. As the lower side view shows, the upper and lower walls of the waveguide look like inductors $L$ (symbolized by coils of "wire"); at the same time, the upper and lower walls form capacitors $C$ (symbolized by parallel plates). The $L$s and $C$s are actually distributed elements; one cannot look at the hollow bar and point to specific $L$s and $C$s because each segment of the waveguide is identical to all the other segments, and each segment represents a minuscule $L$ and $C$. However, it is convenient to regard the $L$s and $C$s as discrete lumped elements that are located at the zero crossover points of the $\underline{E}$ and $\underline{H}$ fields shown in our imaginary "photograph" at this instant of time. The "photograph" of Fig. 4-1 actually depicts four parameters: $V$ (voltage), $I$ (current), $\underline{E}$ (electric field), and $\underline{H}$ (magnetic field).

The expression "polarized light" has become quite commonplace, but polarization applies to all EMFs, not only to light waves. The direction of polarization is, by definition, the same as the direction of the $\underline{E}$ field. In Fig. 4-1, therefore, the polarization of the EMF is vertical. The direction of polarization is one of the important characteristics of an EMF and, by extension, of a photon.

When the EMF emerges into vacuum (or air) at the right end of the waveguide, the $\underline{E}$ and $\underline{H}$ lines remain mutually perpendicular to each other and to the direction of propagation. At the edges, however, unrestrained by the waveguide walls, the EMF beam laterally spreads out (known as diffraction). Because $\underline{E}$ and $\underline{H}$ lines have to be continuous, what happens to the ends of the fields after they leave the waveguide? The positive ends of one $\underline{E}$ line join up with the negative ends of the adjacent $\underline{E}$ lines to form an ever-expanding zigzag pattern as the beam spreads out. The same joining-up of the ends of adjacent lines occurs for the $\underline{H}$ field diffraction.

Returning to Table 3-1, the characteristics of EMF propagation in three representative mediums are listed: vacuum or air, ruby mica, and water (distilled). What EMF characteristics are analogous to the Young's modulus and the density that are listed for sound transmission? The answer is found in the $L$s and $C$s of the waveguide. Inductance is associated with opposition to changing current, and is analogous to the inertia of a mass. The magnetic permeability, $\mu$, is analogous to density of a mass, $\rho_D$. Capacitance is associated with ease of change, analogous to elasticity. The medium's permittivity, $\varepsilon$, is analogous to the reciprocal of the stiffness $(1/Y_0)$, so $\varepsilon$ is analogous to mechanical compressibility.

## Propagation of Photons

A photon is a wave packet (as depicted in Fig. 4-2), an oscillation that carries a minuscule amount of energy

$$E = hf, \quad (4\text{-}1)$$

where $h$ = Planck's constant, and
$f$ = frequency of the photon.

A photon can travel for billions of years through space before it strikes a material object, which causes the photon's energy to be converted into heat, chemical energy, and other forms of energy. Remarkably, regardless of its source or frequency, the photon travels at the speed of light. There is no way that the photon wiggle can propagate, in this manner, unless it is carried by the aether since a vacuum, empty space, cannot sustain that wave motion; Maxwell and his contemporaries viewed this as sufficient proof. Furthermore, the aether is a perfectly elastic medium; the photon can travel, without attenuation due to friction, from one side of the universe to the other.

The above discussion about the energy of a photon ties in with an EMF signal because electric and magnetic fields are forms of energy. Aside from its minuscule magnitude, the main difference between a photon and the EMF shown in Fig. 4-1 is that the EMF is a steady-state sinusoidal signal; its energy is given in joules per second (which is a measure of the power in watts). The energy of a photon is given in joules since the energy of the photon is contained in a well-defined wave packet. The EMF in Fig. 4-1 has a certain power density in watts per square meter. This is the *total* power density carried through the waveguide by the EMF; we do not say that half of this is due to the $\underline{E}$ field and the other half to the $\underline{H}$ field, because $\underline{E}$ and $\underline{H}$ are inseparable. (The magnetic field of the Earth exists without an $\underline{E}$ field, and the electric field of a battery exists without an $\underline{H}$ field, but these fields are not propagating. Fig. 4-1 illustrates a signal, like the one leaving a radio transmitter, that is traveling at the speed of light and consists of inseparably intertwined electric and magnetic fields.)

Photons ignore each other; they do not interact in the same way that electrons do, for example, since electrons repel each other because of their similar (minus) charges. But photons *do* interact in a different sense. A single, isolated photon reveals itself by tiny $\underline{E}$ and $\underline{H}$ lines. (Individual lines do not actually exist, of course, but they are a very convenient manmade concept for visualization and design.) In a laser beam, the edges of one photon's $\underline{E}$ and $\underline{H}$ lines join up with the next, and many $\underline{E}$ and $\underline{H}$ lines coincide. The net effect is that of a huge universe of $\underline{E}$ and $\underline{H}$ lines that combine to form the waveguide fields shown in Fig. 4-1.

# CHAPTER 5

▼

# THE EARLY SEARCH FOR THE AETHER

It is easy enough to detect a sound-wave carrier: place a buzzer inside a jar and start to pump out the air. The loudness of the buzzer gradually diminishes until, with sufficient vacuum, it is no longer heard. Unfortunately, it is not possible to pump the aether out of a jar. It is assumed here that the aether consists of aether particles, EPs, that are the same size or smaller than an electron, and that they occupy the "empty space" inside and between atomic structures.

In this chapter, let's assume that the medium through which the waves are propagating is a "vacuum," and not a material substance such as glass. Also, let's use the more convenient value of $c = 3 \times 10^8$ m/s for the velocity of a photon.

Following Maxwell's EMF revelations, valiant efforts were made to detect the putative aether. The most famous experiment was carried out by Albert A. Michelson and Edward W. Morley in 1887 [12]. It was generally believed that the aether drifts through space without much regard for material objects, such as air, that happen to occupy that space. (This belief was amazingly justified, in 1911, when Ernest Rutherford showed that "solid matter" consists of nuclei that are very far apart, allowing plenty of room for aether particles to drift through a material object.) It was known that the speed of the Earth around the sun is $3 \times 10^4$ m/s. Therefore, the velocity of light should be $c = 3 \times 10^8$ m/s + $3 \times 10^4$ m/s (an increase by a factor of 0.0001) if the ray of light is moving "downstream," and it should be decreased by a factor of 0.0001 if the light is moving "upstream." Well, Michelson and Morley were not able to detect an appreciable, significant, or reliable difference between the speeds of light rays moving upstream and those moving downstream!

The experiment has been repeated many times since 1887. Around 1925, Dayton Miller et al. repeated the Michelson-Morley experiments using more accurate equipment. The conclusions are controversial. There were indications of an aether drift, but this may have been caused by "noise" because the effect is relatively small. In 1933, Miller reported a drift of some 0.2 million m/s (as compared with light at 300 million m/s) toward the Swordfish constellation [2]. (In Fig. 1-1 of this book, the drift is simply shown as a background effect.) A table summarizing the early stages of the search for the aether can be found on page 235 of *Classical Electricity and Magnetism* by W. K. H. Panofsky and M. Phillips [19].

Two possible explanations were offered to explain the results of the search for the aether.

1.  There is no aether. Somehow, photons can propagate for billions of years, through the vastness of the universe, without a carrier, at a velocity independent of photon frequency, and without attenuation due to friction.

2.  There *is* an aether that permeates all of space, but its local component is stationary relative to the Earth. Perhaps it is gravitationally attracted to the Earth, like the Earth's atmosphere of air (or, perhaps, like dark matter). This possibility is depicted in Fig. 1-1. The Earth is labeled *US*, and is pictured as being a "stationary" platform. (This is not quite true; there are small centrifugal accelerations because of the Earth's rotation around the sun plus its daily rotation around its axis. For the purpose of the present discussion, however, we can ignore these accelerations and regard the Earth as a non-accelerating platform.)

### Aether "Atmospheres"

The aether "atmosphere" found in Fig. 1-1 is shown as a finite layer with a sharp motion discontinuity; that is, inside the "atmosphere," the EPs are moving with the Earth; outside, they constitute the aether background. Actually, the motion of the "atmosphere" layer must attenuate exponentially in some fashion, gradually blending in with the background motion. This is accomplished without a change in EP density (the number of EPs per cubic meter). Density change would indicate a change in $c$, different from $3 \times 10^8$ m/s. The fact that $c$ remains constant, to the best of our knowledge, testifies to a constant aether density *but not necessarily constant motion.* Tentatively, however, it is convenient to draw, as well as to think about, a layer of aether "atmosphere" that has a sharp motion discontinuity.

Far off to the right in Fig. 1-1 is another planet, labeled *THEM*, which, for convenience in drawing, is the same size as the Earth. It is speeding away, *relative to US*, with a velocity *v*. It is also a non-accelerating platform, and it is carrying, of course, its own aether "atmosphere."

Between *US* and *THEM* is interplanetary space, with its own aether particles moving or drifting, say, in a northerly direction at some unspecified velocity relative to *US*. Herein resides a strong argument, however, against the model of Fig. 1-1(a): if a laser beam (the photon path) leaves *US* and is directed to *THEM*, it has to bend when it encounters the motion discontinuity. First it bends in an upward direction as it leaves the Earth's aether atmosphere; then it bends downward when it enters the distant planet's aether atmosphere. *These effects have not been observed.* The aberration of starlight, when photons from a distant star enter the Earth's atmosphere, shows that no bending occurs [19] (p. 237). (This is discussed in Chapter 7. The aberration is caused by the Earth's motion around the sun, which results in telescopic changes by an angle of arctan 0.0001, and is based on the velocity of the Earth relative to the velocity of light.)

To summarize, the Michelson-Morley results can be explained if EPs are gravitationally attracted to large, massive bodies, exactly as air molecules are gravitationally attracted to the Earth. When a light beam encounters a transparent material such as glass, it bends because its velocity changes; there is an interaction between the EPs and glass molecules. But bending of light beams entering or leaving the Earth's aether atmosphere has not been observed.

This can be explained, however, by a simple conjecture: the only fact we know about the EPs is that they transmit EMF waves at a velocity of $c = 3 \times 10^8$ m/s. Although the characteristics of a sound wave offer some helpful hints, they are completely different from EMF transmission in one important respect: a sound wave is longitudinal; that is, the molecules oscillate in the same direction as the propagation of the wave. An EMF, however, is transverse; the electric and magnetic fields oscillate at right angles to the direction of propagation. The notion that bending should occur, in Fig. 1-1(a), is a throwback to sound-wave ideology. Aether particles undoubtedly transmit in a completely different way: perhaps they spin, and the spin is somehow transmitted. We have no idea how electric and magnetic fields are transmitted from one aether particle to the next. The direction and speed of spin rotation could be the physical embodiment of an electric and magnetic field.

My conjecture is that a light ray *does not bend* when it reaches an aether motion discontinuity. This is depicted in Fig. 1-1(b). The "photon path" follows its original direction, continuing on at $c = 3 \times 10^8$ m/s, ignoring motion discontinuities.

If Fig. 1-1(b) is correct, it would explain why it has not been possible to detect the aether. Also, if the aether does not do anything in the sense given here, it explains why Einstein avoided it. As Peter Galison wrote in his fascinating and informative book, *Einstein's Clocks, Poincaré's Maps* [15] (p. 324), "Earth's motion through the aether could not be detected...and, *therefore*, so the argument went, Einstein concluded that the aether was 'superfluous.'" Apparently, Albert Einstein was happy that his space-time equations were correct; he had more interesting and important projects on his agenda than trying to figure out how an EMF propagates, so he abandoned the aether. Nevertheless, Henri Poincaré and many other scientists did not regard the aether as "superfluous." Without a reasonable explanation for how an EMF can propagate in a vacuum, the aether hypothesis could never be laid to rest.

This introduces us to a very ironic situation, because Einstein's special relativity is based on the aether! This thesis is explored in Chapter 8.

# CHAPTER 6

▼

# MICHELSON-MORLEY AND MODERN INTERFEROMETERS

The Michelson-Morley (M-M) interferometer is worth looking at because interferometers are extremely important in astronomy and, in general, in operations involving electromagnetic waves [19] (p. 233). There are two requirements here: the signals being processed must have a fixed frequency (monochromatic), and the equipment must combine two signals so that they tend to add (constructive interference) or subtract (destructive interference), depending on the relative phase angles of the two signals.

The M-M interferometer is depicted in Fig. 6-1. We start with a monochromatic beam directed to a half-silvered mirror (HSM). The latter reflects half of the light striking the mirror (ideally), and lets the other half go through the mirror as if it is a sheet of plain glass. (Most of the lines in the figure are drawn at a slant for the sake of clarity even though all of the rays of light should be drawn as either horizontal or vertical lines.)

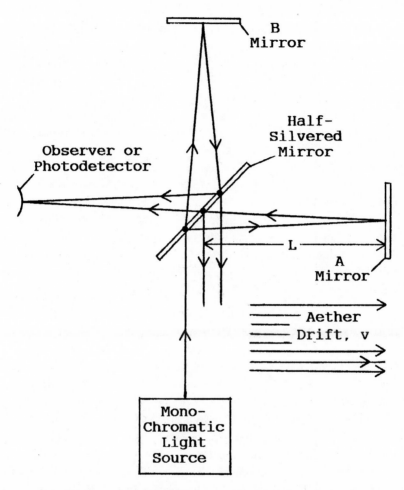

**Fig. 6-1.** The Michelson-Morley interferometer. (Some of the lines are drawn at a slant for the sake of clarity even though all of the rays of light should be drawn as either horizontal or vertical lines.) The aether drift is moving to the right. As the equipment is rotated by 90°, the observer should see a change in the fringe pattern. Michelson and Morley were not able to detect an appreciable, significant, and reliable change.

When the incoming light source strikes the HSM, a horizontal reflection goes to the right a distance $L$, and then is reflected to the left by mirror A. When the reflection from mirror A strikes the HSM, half of the light goes through the HSM and onward to the observer.

When the light source strikes the HSM, half of the light goes through to mirror B, is reflected toward the bottom of the page, strikes the HSM, and is reflected to the left toward the observer.

What will the observer see? The observer will see anything between a relatively bright output signal (if A and B are in phase) to a relatively dark output signal (if A and B are 180° out of phase). Usually, one of the mirrors is slightly tilted so that light-and-dark lines, "fringes," appear. The observer is looking for a *change*.

In Fig. 6-1, the aether particles are drifting to the right with velocity $v$. As the equipment is rotated 90° (so that the aether drift becomes parallel to the light rays from mirror B instead of the light rays from mirror A), the fringe pattern should change. Much to their disappointment, Michelson and Morley were not able to detect a significant change in the fringe pattern when they rotated the interferometer. Richard Milton has the following to say on his Web site, "Aether Drift?" [20]:

> [I]n 1887 a new instrument was built in the basement of Western Reserve's Adelbert College in Cleveland. The instrument consisted of a 1.5-meter square block of sandstone on which was mounted the optical apparatus. The stone block rested on a wooden disk floating on mercury in a cast iron tank. This made it possible to rotate the instrument through 360 degrees with virtually no vibration. At each end of the arms of the apparatus were four mirrors to reflect the light beam back and forth repeatedly, thus effectively extending the path length for the beam to a total of 11 meters.
>
> If the theory of a stationary aether through which the earth moved were correct, then pointing one arm of the apparatus in the direction of the earth's travel through space would produce a displacement of the observed pattern of interference fringes amounting to 0.4 of a fringe width. The apparatus was sensitive enough to detect a displacement much smaller than this amount.

It is instructive to examine the equations involved here. The time taken for the light beam to travel from the HSM to mirror A is

$$t_{HSM-A} = L/(c + v) \quad (6\text{-}1)$$

because the beam is headed downstream, aided by the aether drift. The time taken for the reflection to travel from mirror A to the HSM is

$$t_{A-HSM} = L/(c - v) \quad (6\text{-}2)$$

because the beam is headed upstream, against the direction of the aether drift. Adding the two times, the total round-trip time for the first light beam to travel from the HSM, to mirror A, and back to the HSM is

$$t_A = (2Lc)/(c^2 - v^2). \quad (6\text{-}3)$$

For $v \ll c$, we get

$$t_A \cong (2L/c)(1 + v^2/c^2) \quad (6\text{-}4)$$

In the meantime, because the instrument is moving to the left with respect to the aether drift, the second light beam travels from the HSM, to mirror B, and back to the HSM along the hypotenuse of a right triangle. The base of the right triangle is $Lv/c$, and the hypotenuse of the right triangle is $[L^2 + (Lv/c)^2]^{1/2}$, so the round-trip time for the second beam is

$$t_B = (2L/c)(1 + v^2/c^2)^{1/2} \quad (6\text{-}5)$$

which, for $v \ll c$, yields

$$t_B \cong (2L/c)(1 + 0.5v^2/c^2). \quad (6\text{-}6)$$

The time difference between the two beams is, evidently,

$$t_A - t_B \cong Lv^2/c^3. \quad (6\text{-}7)$$

With a light beam of frequency $f$, the number of equivalent light cycles, $n$, in the time difference is given by

$$n \cong Lfv^2/c^3. \quad (6\text{-}8)$$

Michelson and Morley expected an appreciable value for $n$ due to the orbital velocity of the Earth around the sun [see Fig. 1-2(c)]. They pictured unimaginably small aether particles as rushing through space at a drift velocity $v = 3 \times 10^4$ m/s (or greater), undetected by human observers. Let's assume that they used monochromatic light near the center of the visual spectrum, $f = 6 \times 10^{14}$ Hz, and that the value of $L$ was 5.5 m. (As Milton reported, a total of 11 meters was achieved by forcing the beams to repeatedly reflect back and forth between mirrors.) Then the predicted value of $n$ can be calculated:

$$n \cong (5.5)(6 \times 10^{14})(3 \times 10^4)^2/(3 \times 10^8)^3 \cong 0.11.$$

This is a perfectly reasonable answer. A change of 0.11 cycle can be easily detected.

## Modern Interferometers

As luck would have it, as time went on, our technical prowess advanced prodigiously. It became obvious that one could do much better than Michelson and Morley or even Dayton Miller et al. Precision measurements were proposed simply to discover if the velocity of light was isotropic; that is, the same in any direction, independent of the Earth's movement due to daily rotation or annual revolution around the sun. The new high-precision element was the maser and,

later on, the laser. These instruments can supply a powerful monochromatic signal whose frequency can be accurately measured.

In 1964, T. S. Jaseja et al. reported on a test of special relativity using infrared
masers [3]. According to special relativity, space is isotropic, so that an electromagnetic field propagates everywhere with the same velocity, $c$. (The interpretation in the present book is that this is true only in the aether "atmosphere" that
surrounds every structure that is massive enough to gravitationally hold the
aether in place.) Jaseja et al. found that, within the accuracy they could muster,
space was indeed isotropic.

Next came a stepwise change in philosophy. In the Michelson-Morley interferometer, a light beam travels back and forth between mirrors. Assuming that an
aether drift is present, the time taken for the light to travel downstream is added
to the time to travel upstream; the net time is *always* larger than the time taken if
there is no aether drift [see Equation (6-3)]. In the new system, introduced by A.
Brillet and J. L. Hall in 1979 [4], the change in the resonant frequency of a *cavity
resonator* is measured. (It is claimed that a change in the speed of light will be
accompanied by a change in the *length* of the resonator. I fail to see how the
length can change; in my opinion, the number of wavelengths carried by the cavity changes, and this directly corresponds to a change in the resonant frequency.)
But a laser is locked into a harmonic of the cavity resonator; the resonant frequency change causes the laser frequency to change. The idea is to thereby trans-
late the number of wavelengths in the resonator into a laser frequency change.
Brillet and Hall claimed that they had achieved "a 4000-fold improvement on the
best previous measurement of Jaseja et al" [4].

In 2003, Peter Wolf et al. set up a simple test using stationary (non-rotating)
equipment [6]. They started with a 100 MHz hydrogen maser and multiplied its
signal by 120 to get a frequency of $12 \times 10^9$ Hz. They beat this against an
$11.933 \times 10^9$ Hz cryogenic sapphire oscillator to get a 67 MHz difference signal. They beat the latter against the 67 MHz output of a synthesizer to get, ideally, DC. Actually, the maser or sapphire oscillator frequency was slightly offset
to give a final beat output of approximately 64 Hz, which was fed to a counter.
The Earth's daily rotation, and annual orbit, had a negligible effect on the final
output frequency.

The main difficulty with the systems described in the references discussed
above ([4] and [6]) is this: if the cavity resonance *does not* change, the laser frequency *can* nevertheless change due to several factors such as intrinsic "noise" due
to plasma movement, temperature drift, and minute mechanical imperfections in
the rotating table (if one is used) on which the parts are mounted. These particular laser frequency changes represent noise, of course, and the noise limits the
accuracy of the apparatus. Heroic steps are therefore taken to stabilize the system.

These steps can correct for temperature changes, but not for mechanical short-comings. The cavity resonators have to be made out of a material that has a low-temperature coefficient of expansion and a low creep.

Several different designs have been used [4, 6, 7, 8, 9], but I prefer to use the block diagram created by M. Consoli and L. Pappalardo (shown in Fig. 6-2) because it is clearer than the others. (Fig. 6-2 is taken from a proposal by Maurizio Consoli and Lorenzo Pappalardo: Fig. 2 in "LIEDER: Laser Interferometer for Aether Drift Experimental Research." Further information is available from consoli@ct.infn.it or Lorenzo.Pappalardo@ct.infn.it.) Only the highlights are discussed below.

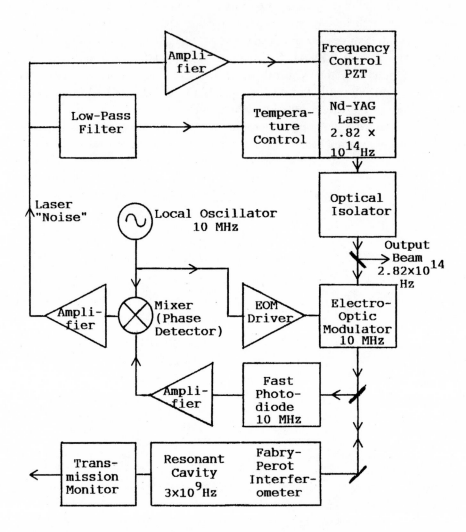

**Fig. 6-2.** Block diagram of modern interferometer used to search for the aether drift. This figure is taken from a proposal by Maurizio Consoli and Lorenzo Pappalardo: Fig. 2 in "LIEDER: Laser Interferometer for Aether Drift Experimental Research." The output beam of this unit normally beats against the output beam of a similar unit, but the two resonant cavities are oriented at right angles to each other.

The resonant cavity, at the bottom of the diagram, is a hollow cylinder with small optical access holes at the ends. It can be constructed from a crystalline sapphire; other possible materials are ULE and Zerodur. Consoli and Pappalardo plan to use a 5-cm long cavity; this is the half-wave length at the fundamental frequency. If the cavity is "filled" with a vacuum and the wavelength is 0.1-meter, the fundamental frequency is $(3 \times 10^8)/(0.1) = 3 \times 10^9$ Hz, as shown in Fig. 6-2. The cavity can also be filled with a dielectric material, with a corresponding change in the fundamental frequency.

Near the upper-right corner of Fig. 6-2 is the Nd-YAG laser whose output frequency is nominally $2.82 \times 10^{14}$ Hz. (This frequency is infrared, near the low-frequency edge of visible light, which begins at $4 \times 10^{14}$ Hz.) The goal is to lock the laser frequency to a suitable harmonic of the cavity resonance. Which harmonic? Dividing $2.82 \times 10^{14}$ by $3 \times 10^9$, the harmonic is calculated to be the 94,000th harmonic! For his contribution to this type of system, John L. Hall shared the 2005 Nobel Prize in Physics [4, 5].

The various elements of the block diagram adhere to what is known as the Pound-Drever-Hall scheme [5]. For further explanation, the reader will have to refer to the paper by Drever et al. and the references found therein because these vital but fine details are beyond the scope of this book [5]. You can get a feel for what is involved, however, from the paper's abstract [5]:

> We describe a new and highly effective optical frequency discriminator and laser stabilization system based on signals reflected from a stable Fabry-Perot reference interferometer. High sensitivity for detection of resonance information is achieved by optical heterodyne detection with sidebands produced by rf phase modulation. Physical, optical, and electronic aspects of this discriminator/laser frequency stabilization system are considered in detail. We show that a high-speed domain exists in which the system responds to the phase (rather than frequency) change of the laser; thus with suitable design the servo loop bandwidth is not limited by the cavity response time. We report diagnostic experiments in which a dye laser and gas laser were independently locked to one stable cavity. Because of the precautions employed, the observed sub-100 Hz beat line width shows that the lasers were this stable. Applications of this system of laser stabilization include precision laser spectroscopy and interferometric gravity-wave detectors.

The two end results of this new design are shown in Fig. 6-2: feedback is available to control the temperature of the laser, and feedback is available to control piezoelectric movement (PZT), which makes adjustment of the laser frequency possible.

The standard procedure for this setup is to beat the laser output against the output of a similar unit, with the two cavities at right angles to each other. Thus, if aether drift is picked up by one of the cavity resonators, the second cavity unit will pick it up six hours later. Disregarding the phase, the beat frequency should peak twice a day. If the equipment is rotated, the beat amplitude should peak at every multiple of 90°.

In 2003, Holger Muller et al. used a setup similar to the one shown in Fig. 6-2 [7]. A similar setup was also used by P. Antonini et al. in 2005 [9].

What are the net results of these investigations into the existence of the aether? It has been shown that space is isotropic: the velocity of light is the same in any direction (with respect to the physics laboratory). This result is in accordance with the theory of special relativity, but this result is also consistent with the conjecture that a stationary aether is held in place by gravitational attraction.

For each cavity, where $f = c/(2L)$, Equation 6-8 becomes

$$n = 0.5(v/c)^2. \quad (6\text{-}9)$$

This is a weak circuit element, whose response increases as the square of the aether drift velocity. As of 2005, the equipment has not been able to reliably identify an aether drift. I can think of at least three ways to increase the sensitivity of the current equipment.

1.  A third cavity resonator could be mounted at right angles to the first two. It is obvious, from Fig. 6-2, that each interferometer unit is complex and expensive; nevertheless, a third unit, with a mutually perpendicular cavity resonator, should be added if the first two show definite promise.

2.  Ideally, a manned or unmanned space vehicle could be used to carry the apparatus as part of its load, enabling the achievement of spectacular velocities.

3.  One can endeavor to measure the velocity of light directly. In principle, this is the essence of simplicity. Place a laser, whose frequency is highly accurate, in the physics laboratory. Then count the number of cycles intercepted each second, say, as a counter recedes from the laser. For example, a laser frequency of $10^{14}$ Hz has a wavelength of $10^8/10^{14} = 10^{-6}$ m, or one million cycles per meter. If the counter recedes from the laser at a speed of 1 meter/second, the intercepted frequency should decrease by $10^6$ Hz; and so forth. This would be a challenging and worthwhile PhD thesis topic.

# CHAPTER 7

▼

# STELLAR ABERRATION
# VERSUS THE AETHER

In 1887, the Michelson-Morley experiment showed that the aether, if it exists, is carried along by the Earth. In other words, every large object (the Earth, the moon, the sun, etc.) is immersed in its own aether and carries its aether along with it in a manner similar to the way in which the Earth carries its atmosphere of air. This depiction may seem to be far-fetched, but remember that dark matter (DM) is gravitationally bound to the stars that are immersed in it; in fact, that is how DM was discovered [21]. It is accordingly conjectured that DM and EPs may be identical.

In 1955, a well-known textbook by W. K. H. Panofsky and M. Phillips went into considerable detail to once again examine the aether hypothesis [19]. Their conclusion was that the aether does not exist.

The aether hypothesis is illustrated in Fig. 1-1. The Earth is surrounded by an aether "atmosphere." The aether "atmosphere" undoubtedly trails off exponentially, but the worst case is depicted: there is a sharp discontinuity between the aether "atmosphere" and interplanetary space. The discontinuity is represented by a cloud labeled "aether drift."

This portrayal of the solar-system's aether is an important bone of contention in the aether hypothesis. It implies (Panofsky and Phillips, p. 231) "that there exists a unique privileged frame of reference, the classical 'aether frame,' in which

Maxwell's equations are valid and in which light is propagated with the velocity $c$" [19]. The background aether is undoubtedly drifting through intergalactic space, and it has to obey Einstein's special relativity equations (which are considered in Chapter 8). With respect to an observer on an inertial platform (the Earth), the component of the drifting aether that is moving toward (or away) from the observer becomes shorter. Also, a clock carried along toward (or away) from the observer becomes slower [10].

These space-time effects are negligible unless the aether is drifting toward (or away) from the Earth at a "relativistic" speed (a speed larger than one-third of the speed of light). (One can also neglect the slight acceleration that the Earth suffers because of its rotation and its orbit around the sun.) The most important reason for abandoning the aether concept is that the "background" aether has never been detected.

Although the large-scale manifestations of the aether have not been uncovered (aside from the fact that a photon can propagate indefinitely, without loss to the medium due to friction), there is plenty of evidence that the aether may be involved in the weird subatomic effects of quantum mechanics [22].

But the raison d'etre of the present chapter is that the aberration of stars has been cited as "proof" that the aether does not exist. Consider Fig. 7-1(a), where a telescope on Earth is lined up with a distant star. In Fig. 7-1(b), at t = 0, a photon from the distant star enters the telescope tube. At t = 0⁺, the photon has progressed halfway down the tube. Because the Earth's orbit carries the telescope to the left, the photon's relative motion is toward the right. At t = 0⁺⁺, the photon strikes the light-sensitive receiver (assumed to be a film in the following discussion) toward the right, as shown. The angular movement is given by arctan 0.0001. [The value 0.0001 is equal to the Earth's speed ($3 \times 10^4$ m/s) divided by the velocity of light ($3 \times 10^8$ m/s).] Although it is minuscule, this amount of aberration is easily detected and measured.

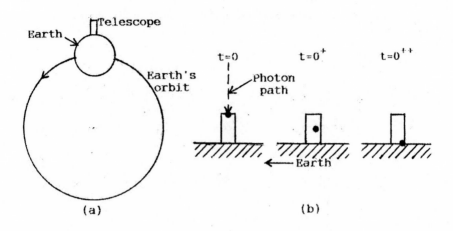

**Fig. 7-1.** The basis for stellar aberration. (a) A telescope on Earth is lined up with a distant star. For the sake of clarity, the sun is omitted. (b) This shows the path of a photon that enters the telescope tube at $t = 0$. Because the Earth's orbit carries the telescope to the left, the photon's relative motion is toward the right. The angular movement is 20.5" (arcseconds).

To round out the aberration phenomenon, Fig. 7-2(a) depicts the telescope six months after the view shown in Fig. 7-1(a). As shown in Fig. 7-2(b), the Earth is moving to the right six months after the view shown in Fig. 7-1(a), so the star's photons drift to the left. If the film is repeatedly exposed for a year whenever the telescope is lined up on the star, a circle of aberration is found; its diameter corresponds to 41" of arc.

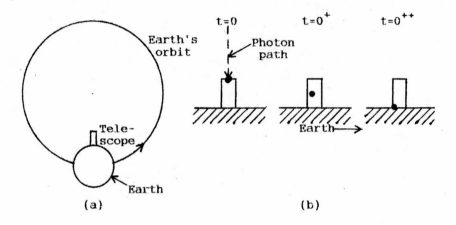

**Fig. 7-2.** Six months after the view shown in Fig. 7-1. (a) The telescope is again lined up with the star. (b) This shows the path of a photon that enters the telescope tube at $t = 0$. Because the Earth's orbit carries the telescope to the right, the photon's relative motion is toward the left.

It has generally been assumed that the Earth's aether, moving with respect to the sun, would grab the incoming photon and steer the photon so that it would go straight down the center of the telescope's tube. This is illustrated in Fig. 7-3(a), where the photon is deflected toward the right. In Fig. 7-2(b), this would cancel the photon's leftward drift, yielding a film exposure free of aberration. The details of the photon's path in Fig. 7-3(a) are described in Fig. 7-3(b). For the sake of clarity, only the aether particles (EPs) involved in the photon's propagation are shown. According to the aether theory conjecture, the spin at t = 0 represents the photon entering the upper left-hand EP. At times t = 2, 4, 6, and so on, the spin of the EPs that are shown correspond to the location of the photon at times t = 2, 4, 6, and so on. (The small arrow is symbolic and is not meant to show the actual spin.)

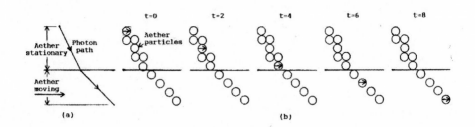

**Fig. 7-3.** Illustrating the general assumption that the Earth's aether, moving with respect to the sun, would grab an incoming photon and steer it in the direction of aether motion. (a) A photon is deflected toward the right. (b) This shows the details of the photon's path. For the sake of clarity, only the aether particles involved in the photon's propagation are shown. The small arrows are symbolic and are not meant to show the actual spins.

Since the characteristics of the aether are not known, it is conjectured in the present chapter that the interpretations of Fig. 7-3 are incorrect; that, in fact, the moving aether *does not* affect the flight path of a photon. This is illustrated in Fig. 7-4. In Fig. 7-4(a), the photon's path is a straight line, despite the movement of the aether in the lower section of the drawing, so that stellar aberration *does* occur despite the Earth's aether atmosphere. The EP spin details are shown in Fig. 7-4(b). The photon paths that are shown are those found in Fig. 7-1 and Fig. 7-2. What is the basis for the claim that the path of a light beam is independent of aether motion? It is that the velocity of light does not change in crossing the motion discontinuity of Fig. 7-4(a). As the spin is transmitted from one EP to the next at the speed of light, it is blind to relative motion (provided that the velocity of light remains constant in going from one region of space to the next).

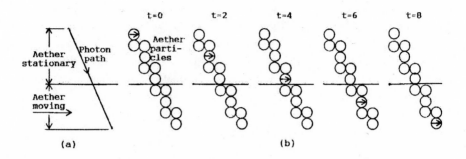

**Fig. 7-4.** Illustrating the conjecture made in the present book that the Earth's aether does not affect the flight path of a photon. (a) The photon's path is a straight line. (b) This shows the details of the photon's path. Only the aether particles involved in the photon's propagation are shown.

This can be contrasted with a beam of light passing from air to glass (or vice versa), as depicted in Fig. 7-5. The diameter of each EP does not change, but the velocity of light is given by $1/(\mu\varepsilon)^{1/2}$, where $\mu$ is magnetic permeability and $\varepsilon$ is electric permittivity. In glass, the permittivity is greater than in air, so the electric field changes and the corresponding spin, in glass, is different from what it is in air. The reduced velocity of light causes the beam to bend when it crosses the permittivity discontinuity.

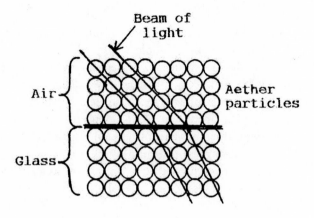

**Fig. 7-5.** Path of a beam of light passing from air to glass (or vice versa). The change in the velocity of light causes the beam to bend when it crosses the permittivity discontinuity.

## Motion-Independence Conjecture

The motion-independence conjecture has far-reaching consequences for the universe. Returning to Fig. 1-1, a beam of light leaving the Earth, crossing the drifting aether field, and crossing the distant planet's aether atmosphere follows a straight-line path. This agrees with the locally constant velocity of light in a non-accelerating medium. (The centrifugal accelerations of the Earth are negligibly small.)

The conjecture is that the universe is a huge cloud of aether. Here and there are minor discontinuities formed by stars, planets, and the like. We cannot detect the turbulence associated with these discontinuities because light (or any EMF) does not bend in encountering a discontinuity unless there is a change in permeability and/or permittivity. (This ignores the slight bending associated with Einstein's space-time curvature.)

Space has no meaning beyond the edge of the aether cloud. Are permeability and permittivity everywhere constant? Probably not, as the cloud expands (or contracts). It is foolhardy to assume that the natural constants, such as the velocity of light, have been and will be invariant over all space and time [23].

# CHAPTER 8

▼

# SPECIAL RELATIVITY

Let's turn the clock back some 100 years, to 1905, when Einstein was twenty-six years old. Maxwell's aether implied that the universe was filled with the aether background of Fig. 1-1, with the aether drifting about relatively slowly (compared to the speed of light) through turbulence created by the stars, planets, and moons. Measurements indicated, whenever they could be made, that the *local* velocity of light is, always, $c = 3 \times 10^8$ m/s. Einstein adopted this as a guiding principle, never to be violated. See "On the Electrodynamics of Moving Bodies," by Albert Einstein [16].

Next, in Fig. 1-1, suppose that the planet at the right was retreating from *US* at one-third the speed of light, or at $1 \times 10^8$ m/s. Here is how I imagine that Einstein would describe a beam of light [the photon path in Fig. 1-1(b)] sent from *US* to *THEM*:

"The beam leaves the Earth traveling at $c = 3 \times 10^8$ m/s. When it encounters interplanetary space, it continues in a straight line at a speed of $c = 3 \times 10^8$ m/s. Eventually, the beam catches up with the $1 \times 10^8$ m/s receding planet's atmosphere. The beam somehow speeds up to $4 \times 10^8$ m/s relative to *US*, which is $3 \times 10^8$ m/s relative to *THEM*. The beam thus lands at the proper speed."

I imagine that Einstein would continue with the following: "Relative to *THEM*, interplanetary space and the Earth (*US*) are retreating to the left at a velocity of $1 \times 10^8$ m/s. Therefore, if the people on *THEM* send a light beam to the people on *US*, it would at first travel to the left at $3 \times 10^8$ m/s relative to

*THEM*. When the beam reaches interplanetary space, it would speed up to $4 \times 10^8$ m/s relative to *THEM*, which is $3 \times 10^8$ m/s relative to *US*. This time the beam lands on Earth at the proper speed."

Today, because the universe is expanding, we know that there really are planets receding from us at a velocity of $1 \times 10^8$ m/s. Suppose, now, that a 100 Hz "light" signal originates at *THEM* and is directed to *US*. When the signal reaches the equivalent of the above interplanetary space, and its velocity increases to $4 \times 10^8$ m/s relative to the receding planet, the frequency of the signal decreases to 75 Hz. This is the "Red Shift" (the ratio is 1.33, or red shift $z = 0.33$).

To Einstein's imagined assessment of the photon path of Fig. 1-1(b), I would only add "Restore the aether" [22]. This would provide the physical basis for a light velocity of $c = 3 \times 10^8$ m/s in the Earth's "aether atmosphere," and the same value at a planet receding from *US* at a velocity of $1 \times 10^8$ m/s. Furthermore, my conjecture is that a sufficiently sensitive apparatus, carried aboard a space vehicle, could detect the movement of the aether drift. At the very least, it should detect the aether movement relative to the space vehicle. Equipment that can detect motion in three mutually perpendicular directions would be useful. This equipment should make an *absolute* measurement rather than a *relative* measurement, as Michelson and Morley did when they looked for a *relative change* as they rotated their apparatus.

In reality, the sharp motion discontinuities of Fig. 1-1 must be rounded off so that all of the changes discussed above are gradual, with one exception: the velocity of light relative to *US* and to *THEM* is, always, $c = 3 \times 10^8$ m/s.

All of this hopping back and forth between $4 \times 10^8$ m/s relative to *US* and $3 \times 10^8$ m/s relative to *THEM*, and vice versa, hides an astonishing fact that Einstein recognized in 1905: the perception of time (and, subsequently, space) for *US* has to be different from what it is for *THEM*! The proof is simple (and here I am borrowing heavily from N. David Mermin's *Space and Time in Special Relativity*) [11]. The proof assumes that the *local* velocity of light is $3 \times 10^8$ m/s.

Shown in Fig. 8-1(a) is a clock constructed by attaching a mirror to the end of a stick that is $\ell$ meters long. At time $t_0 = 0$, we launch a pulse of light; it strikes the mirror and is reflected back to a detector, reaching it at $t_1$ seconds. From RT = D, we get

$$ct_1 = 2\ell. \quad (8\text{-}1)$$

Now, the people on *THEM* have an identical timepiece, so they also get $ct_1 = 2\ell$. But to *US*, looking through telescopes, the clock on *THEM* is seen as depicted in Fig. 8-1(b). Three views are shown.

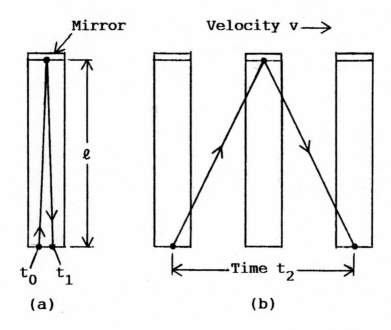

**Fig. 8-1.** A clock that demonstrates slower time, relative to *US*, on a rapidly receding planet *THEM*. (a) The clock consists of a light-pulse generator at $t_0$, a mirror, and a detector at $t_1$. (b) This is an identical clock, as it is seen through an imaginary telescope by an observer on Earth. The three views show, respectively, the light-pulse starting off, arriving at the mirror, and arriving at the detector.

In the first view, the pulse of light is just starting off. Because the clock is moving to the right with velocity $v$ (as seen by *US*), the light beam takes a slanting path to the right. In view 2, it strikes the mirror. In view 3, it is reflected back to the detector. As seen by *US*, the velocity of the light beam is $c = 3 \times 10^8$ m/s, but its path is the hypotenuse $d$ of two identical right triangles: their height is $e$ and their base is $vt_2/2$, so that

$$d = [e^2 + (vt_2/2)^2]^{1/2} = ct_2/2. \quad (8\text{-}2)$$

Eliminating $e$ in Equations 8-1 and 8-2, we easily get

$$t_2/t_1 = 1/[1 - (v/c)^2]^{1/2}. \quad (8\text{-}3)$$

In this equation, $t_2$ is always greater than $t_1$, so we perceive that the *THEM* clock is slow. Some numerical values are presented in Table 8-1. In the previous example, where $v = c/3$, we get $t_2 = 1.061t_1$, or that the clock on *THEM* runs slow by a factor of 0.943.

**Table 8-1.** Perception by *US* of how slow the clocks on *THEM* are as a function of velocity, $v$.

| $v \times 10^8$ m/s | $v/c$ | $t_2/t_1$ | $t_1/t_2$ |
|---|---|---|---|
| 0 | 0 | 1 | 1 |
| 0.3 | 0.1 | 1.005 | 0.995 |
| 1 | 0.333 | 1.061 | 0.943 |
| 1.8 | 0.6 | 1.25 | 0.8 |
| 2.4 | 0.8 | 1.667 | 0.6 |
| 3 | 1 | $\infty$ | 0 |

Since $v$ is squared in Equation 8-3, the clock on *THEM* also runs slow if the planet is approaching *US*, in which case $v$ is negative [reverse the arrows in Fig. 8-1(b)]. Therefore, on any planet that is receding from *or* approaching *US*, the clocks run slow relative to the clocks on planet *US*. Most amazing of all is that *all* biological and time processes run slow, synchronized with the slow clocks, so that people age more slowly relative to *US*.

This is also true in reverse. Relative to *THEM*, planet *US* is receding with velocity $v$. Therefore, while their clocks keep perfect time, they perceive that the clocks on planet *US* run slow.

## Unequal Aging Example

If a spaceship departs from *US*, and subsequently returns to *US*, will the people aboard the spaceship return younger than *US*? Here, "vice versa" is not valid because, in order to return, the spaceship has to undergo a tremendous midcourse deceleration and reacceleration. To properly answer such questions, one should plot a Minkowski diagram, such as the one found in Fig. 8-2.

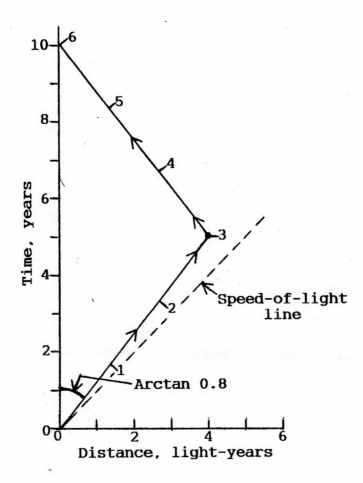

**Fig. 8-2.** The Minkowski diagram for a spaceship receding from *US* at a speed of 0.8*c* = 2.4 × 10⁸ m/s for 5 years (3 years for *THEM*) followed by spaceship's return at a speed of 0.8*c*. The spaceship passengers disembark 4 years younger than *US* inmates. Planet *US* is stationary (distance = 0) along the vertical axis. With the scales shown, the speed of light is represented by a 45° line.

Fig. 8-2 is a Minkowski plot for a spaceship (*THEM*) that travels away from *US* at a velocity $2.4 \times 10^8$ m/s (that is, at 0.8 times the speed of light). After 3 years of *THEM* time, the ship turns around and heads for *US*, again at a velocity of $0.8c$. The diagram is a plot of time versus distance, but time is given in years and the distance is listed in light-years. With the scales shown, the speed of light is represented by a 45° line. The *THEM* locus starts out at an angle of arctan $(0.8) = 38.7°$. According to Table 8-1, $t_2/t_1 = 1.667$, so 3 years on *THEM* shows up as the same time (vertical scale) as 5 years on *US*. The voyage ends with 6 years on *THEM* being equal to 10 years on *US*; that is, the spaceship people arrive 4 years younger than the inmates of *US* (which could be a problem for the IRS), but they had to survive that terrible midcourse reversal of direction.

The Minkowski diagram can reveal much more. Fig. 8-3 is a plot of the above voyage with light pulses broadcast from the people on *US* at 1-year intervals (solid lines) while the people on *THEM* are sending similar light pulses (dashed lines). The light pulses from *US* are 45° lines with a positive slope; the first pulse, sent at $t = 1$ year, arrives at *THEM* at their 3-year point. Subsequent pulses from *US* arrive during the *THEM* return trip, 3 times a year. The light pulses from *THEM* are 45° lines with a negative slope. The first pulse, sent at $t = 1$ year, arrives at *US* at our 3-year point. Their 3-year pulse arrives at our 9-year point. Subsequent pulses arrive 4 months apart.

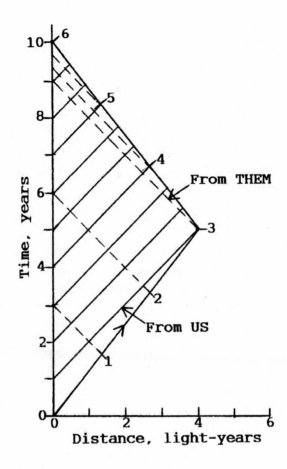

**Fig. 8-3.** The diagram of Fig. 8-2 if we transmit a light-beam pulse every year (solid lines) to *THEM* while they, likewise, transmit a light-beam pulse to *US* every year (dashed lines). The lines from *THEM* have a −45° slope.

All of the above discussion about a spaceship zooming along at $2.4 \times 10^8$ m/s is academic because of the tremendous amount of energy required to accelerate a vehicle to this velocity. The spaceship has to carry its own fuel, of course. Einstein's special relativity has been verified using the atomic equivalents of "space flight," notably with muons. Muons are produced in the Earth's upper atmosphere during collisions between cosmic ray particles and air molecules. In a laboratory on Earth, muons decay in 2.2 microseconds, on average. The "clock" of a high-speed muon that is headed toward the measuring equipment, on the surface of the Earth, is slowed down enough for it to reach ground level before it decays [12] (p. 85).

## Unequal Length Example

Finally, consider how relative velocity causes a perceived change in space (actually, a reduction in length), also known as the Lorentz Contraction (after Hendrik A. Lorentz). This contraction is drawn to scale, in Fig. 8-4, if the velocity of planet *THEM* is $0.6c = 1.8 \times 10^8$ m/s.

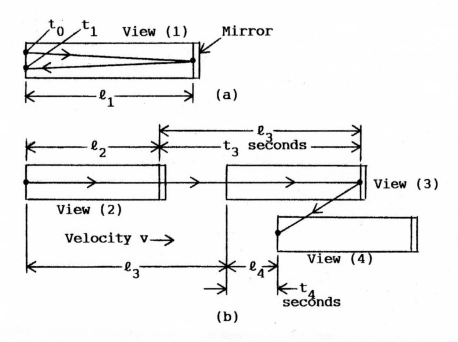

**Fig. 8-4.** The clock of Fig. 8-1 with its orientation changed in order to demonstrate the shortening of sticks (length), relative to *US*, on the rapidly receding planet *THEM*. The velocity of planet *THEM* in this diagram is $v = 0.6c = 1.8 \times 10^8$ m/s. (a) This shows the clock on the stationary planet, *US*. (b) This shows an identical clock, as it is seen through an imaginary telescope by an observer on Earth. Views (2), (3), and (4) demonstrate, respectively, the light pulse starting off, arriving at the mirror, and arriving at the detector. [View (4) actually overlaps View (3), so it is drawn below View (3) for the sake of clarity.]

In Fig. 8-4(a) [also called View (1)], we again have a clock constructed by attaching a mirror to the end of a stick that is $\ell_1$ meters long. At $t_0 = 0$, we launch a pulse of light; it strikes the mirror and is reflected back to a detector, reaching it at $t_1$ seconds. We get

$$ct_1 = 2\ell_1. \quad (8\text{-}4)$$

The people on planet *THEM* have an identical timepiece, so they also get $ct_1 = 2\ell_1$. But to *US*, looking through telescopes, the clock on planet *THEM* is seen as depicted in Fig. 8-4(b). Three views are shown.

In View (2), the pulse of light is just starting off. Because the clock is moving to the right with velocity $v$ (as seen by *US*), the light beam has to travel a considerable distance before, in View (3), it strikes the mirror. In View (4), it is reflected back to the detector. [Because View (4) actually overlaps View (3), it is shown below View (3) for the sake of clarity.]

As always, as seen by *US*, the velocity of the light beam is $c = 3 \times 10^8$ m/s. Then

$$ct_3 = \ell_2 + \ell_3 \quad (8\text{-}5)$$

and

$$vt_3 = \ell_3, \quad (8\text{-}6)$$

where $t_3$ is the time between Views (2) and (3),

$\ell_2$ is the length of the stick as perceived by *US*, and

$\ell_3$ is the distance the mirror moves in $t_3$ seconds.

From Equations 8-5 and 8-6, eliminating $\ell_3$, we get

$$t_3(c - v) = \ell_2. \quad (8\text{-}7)$$

Similarly, we get

$$ct_4 = \ell_2 - \ell_4 \quad (8\text{-}8)$$

and

$$vt_4 = \ell_4, \quad (8\text{-}9)$$

where $t_4$ is the time between Views (3) and (4) and

$\ell_4$ is the distance the mirror moves in $t_4$ seconds.

From Equations 8-8 and 8-9, eliminating $\ell_4$, we get

$$t_4(c + v) = \ell_2. \quad (8\text{-}10)$$

The next step is to use the perceived slowing down of the clocks on planet *THEM*, as given by Equation 8-3. In Fig. 8-4, the total perceived time, $t_3 + t_4$, takes the place of $t_2$ in Equation 8-3. This gives us the following:

$$t_1/(t_3 + t_4) = [1 - (v/c)^2]^{1/2}. \quad (8\text{-}11)$$

Finally, we are interested in length rather than time. Substitute for $t_1$, $t_3$, and $t_4$ from Equations 8-4, 8-7, and 8-10 to get

$$\ell_2/\ell_1 = [1 - (v/c)^2]^{1/2}. \quad (8\text{-}12)$$

In this equation, $e_2$ is always less than $e_1$, so we perceive that the stick on planet *THEM* has shortened. The numerical values of Table 8-1 are again applicable if the last column stands for $e_2/e_1$.

The numerical values used in drawing Fig. 8-4 are the following: $v = 0.6c$, $e_1 = 10$, $e_2 = 8$, $e_3 = 12$, $e_4 = 3$, $t_1 = 20$, $t_3 = 20$, and $t_4 = 5$.

Since $v$ is squared in Equation 8-12, the sticks on *THEM* also shorten if the planet is approaching *US*, in which case $v$ is negative [reverse the arrows in Fig. 8-4(b)]. Therefore, on any planet that is receding from *or* approaching *US*, we see a shortening or flattening of material objects, but only in the direction that the planet is moving away from or toward *US*. In Fig. 8-2, when the spaceship inmates disembark after 10 *US* years (or 6 *THEM* years), will their faces be flattened? Definitely not! As the spaceship decelerates from $v = 0.8c = 2.4 \times 10^8$ m/s to 0 m/s, the flattening (as seen in our telescopes) will gradually vanish. But they *will* be 4 years younger than the people on planet *US*.

## Summary

The basic ideas behind special relativity are simple and are illustrated in Fig. 1-1. Because all motions in Fig. 1-1 are given relative to the Earth, the latter is pictured as being "stationary." Not surprisingly, in the atmosphere surrounding Earth, the velocity of light is, everywhere, $3 \times 10^8$ m/s relative to *US*.

Far off to the right is a planet, *THEM*, that is flying away from Earth at tremendous speed. For convenience, the planet *THEM* is shown as being the same size as Earth. Not surprisingly, in the atmosphere surrounding *THEM*, the velocity of light is, everywhere, $3 \times 10^8$ m/s relative to *THEM*.

Here is a textbook definition of special relativity [11] (p. 7): "The laws of physics—including the behavior of light—must be exactly the same for any two observers moving with constant velocity relative to each other." This leads to some universe-shaking conclusions, such as 1) clocks on planet *THEM* run slow relative to clocks on planet *US* (but normal, of course, to *THEM*), and 2) objects on planet *THEM* shorten, relative to *US*, in the direction that planet *THEM* is moving toward *US*.

Why do we care if the clocks on planet *THEM* run slow and the people on planet *THEM* are flattened relative to *US*? Because these effects imply two things: 1) that the time on planet *THEM* may be different from the time on planet *US* (allowing people on planet *THEM* to appear to age more slowly than people on planet *US*), and 2) that space can be curved because you cannot flatten a three-dimensional object without adding curves to some of the surfaces. Furthermore, two events that appear to be simultaneous to an observer on Earth may not appear to be simultaneous to an observer on planet *THEM*.

Since the Earth is flying away at tremendous speed relative to *THEM*, it appears to *THEM* that Earth clocks are slow and Earth people are flattened. (The unequal-aging scenario requires unilateral acceleration.)

# CHAPTER 9

▼

# AN AETHER PARTICLE (EP)

Time has no beginning, or at least none that makes sense to us. Similarly, space has no beginning. Can we cut an electron up into smaller and smaller pieces? But in visualizing the aether, we have to *start somewhere* to anchor our discussion to, hopefully, the real world around us. I choose to start with an aether that consists of "particles." Here an attempt is made to "visualize" an aether particle given that the main raison d'etre of an EP is to carry electromagnetic waves. *The following description is completely conjectural.*

A sound wave propagates longitudinally as one molecule "bumps into" its neighbor (or leaves a hole that a neighbor can fill). A plane electromagnetic wave propagates transversely; the electric ($\underline{E}$) and magnetic ($\underline{H}$) fields are at right angles to each other and to the direction of propagation. Transmission of aether particles occurs by means of spin. The spin of an EP is *somehow* picked up by its neighbor. In the physical embodiment of these ideas, in Fig. 9-1, an EP is a spherical, spinning body.

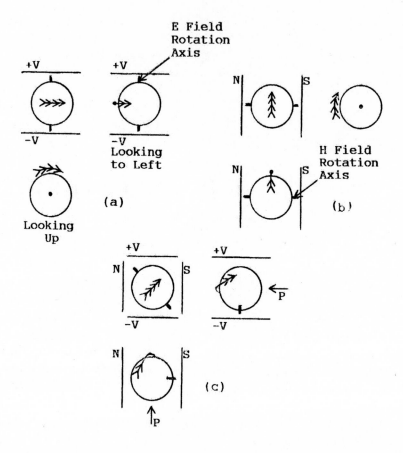

**Fig. 9-1.** The conjectured aether particle (EP). It is a sphere whose diameter is $0.18 \times 10^{-15}$ m and whose mass is $7.5 \times 10^{-68}$ kg. (a) This shows an EP in an electric field. The spin number of "revolutions per second" is proportional to the field intensity. (b) This shows an EP in a magnetic field. The spin is at right angles to the spin shown in (a). (c) This shows an EP in an electromagnetic field that is propagating in the P direction.

The two specifications we would most like to know about an EP are its size and its weight.

The size of an EP is undoubtedly very tiny, having evaded discovery until now. A reasonable guess is that an EP is the size of an electron, but this guess suffers from the fact that we do not really know the diameter of an electron. The "classical electron diameter" is $2.8 \times 10^{-15}$ m = 2.8 femtometers [17] (p. 312), but this is unreasonably high because it is comparable to the diameter of a small nucleus. A calculation follows that gives a sensible answer.

Assume that the density of an electron is equal to that of a neutron or proton. A uranium$^{238}$ nucleus has a diameter of 13.6 fm and contains 238 nucleons (neutrons and protons). Neutrons and protons each have a mass of around $1.67 \times 10^{-27}$ kg. Assuming that the neutrons and protons are densely packed, we get a density of $3.025 \times 10^{17}$ kg/m$^3$. Given that the mass of an electron is $9.109 \times 10^{-31}$ kg, its diameter turns out to be $0.18 \times 10^{-15}$ m = 0.18 fm. This is reasonable given that the diameter of the U$^{238}$ nucleus is 13.6 fm.

Moving on to the weight of an aether particle, a reasonable guess is that the aether is the same as dark matter or, at least, that the weights of aether particles and dark matter particles are equal. In accordance with this proposal, let's find the density of dark matter (DM).

There is no exact model for dark matter. Here it is assumed that DM is a "cloud" that is uniformly distributed inside a sphere whose diameter is that of our Milky Way galaxy. The calculation of the weight of an aether particle is presented below.

The number of neutrons plus the number of protons plus the number of electrons in the universe is equal to $10^{80}$ particles (based on various sources).

Assuming that two-thirds of these particles are neutrons or protons, and ignoring the mass of electrons, the mass of the universe is $(10^{80}) \times (1.67 \times 10^{-27}$ kg$) \times (2/3) = 10^{53}$ kg.

There are $10^{11}$ galaxies. This means that the mass of a typical galaxy, such as the Milky Way, is $10^{42}$ kg.

The mass of DM in the Milky Way is ten times this, or $10^{43}$ kg.

The radius of the Milky Way is 60,000 light-years, and a light-year is equal to $9.46 \times 10^{15}$ m, so the radius of the Milky Way is $5.67 \times 10^{20}$ m.

The volume of a sphere is $4\pi r^3/3$, so the volume of the DM sphere is $7.65 \times 10^{62}$ m$^3$.

This means that the density of dark matter is $1.3 \times 10^{-20}$ kg/m$^3$. Although this is incredibly small, it is far from zero.

How many aether particles are there in a cubic meter? Assuming that an EP is a cube $0.18 \times 10^{-15}$ m on a side, there are $1.71 \times 10^{47}$ EPs/m$^3$.

Finally, the mass of a single EP is $(1.3 \times 10^{-20})/(1.71 \times 10^{47}) = 7.5 \times 10^{-68}$ kg. For comparison, the mass of an electron is $9.109 \times 10^{-31}$ kg. Despite the assumption that DM is ten times as heavy as ordinary matter, when the DM cloud is spread over the entire galaxy, the mass per EP is so small that it probably cannot be detected by present-day instruments.

The direction of spin depends upon whether the EP is in an $\underline{E}$ field, or an $\underline{H}$ field, or both. Because the EP has mass, its spin represents kinetic energy; in fact, it *is* the energy of the field. For example, the energy of an electric field in vacuum is $(\varepsilon_0 \underline{E}^2)/2$, where $\varepsilon_0$ is the permittivity. The energy increases as the square of the field intensity; this exactly correlates with the kinetic energy of a moving mass, $(mv^2)/2$. The velocity (or, in Fig. 9-1, spin revolutions per second) is proportional to the field intensity.

Fig. 9-1(a) depicts an EP in an electric field between the $+V$ and $-V$ plates. In the view "looking up," it is assumed that the spin is clockwise. This is the physical embodiment of an electric field.

Fig. 9-1(b) shows an EP in a magnetic field between the N and S poles. The spin is at right angles to the electric-field spin. In the view "looking to the left," it is assumed that the spin is clockwise. This is the physical embodiment of a magnetic field.

Finally, Fig. 9-1(c) shows an aether particle in a plane electromagnetic wave. The EP simultaneously gets electric and magnetic fields that are at right angles to each other. The resulting spin rotation axis is at an angle of $45°$. This is a field that is propagating as spin is transmitted from one EP to the next. In the upper-left-hand view, energy is propagating out of the sheet of paper; in the view "looking to the left," this is represented by the arrow labeled P. Similarly, in the view "looking up," propagation is represented by the arrow labeled P.

The velocity of light in a vacuum is given by $c = 1/(\mu_0 \varepsilon_0)^{1/2}$, where $\mu_0$ is the permeability. The measured values are $\mu_0 = 4\pi \times 10^{-7}$ henries/meter and $\varepsilon_0 = 8.854 \times 10^{-12}$ farads/meter. In air, as the pressure changes, the molecules don't change, but, as they move closer together or farther apart, the velocity of sound changes. Is this also true for the velocity of light? If so, as the universe expands, and the aether particles move farther apart, the velocity of light can change (increase or decrease) as $\mu_0$ and $\varepsilon_0$ change. This, needless to say, has mind-boggling implications for cosmology [23].

### Electron versus Aether Particle

A few words are in order regarding the flight of an electron through a "vacuum" populated by EPs. An electron has a mass of $9.1 \times 10^{-31}$ kg while an EP has a mass of $7.5 \times 10^{-68}$ kg. Therefore, the electron is around $10^{37}$ times as heavy as an EP; this is a huge value. Nevertheless, for example, an electron that is attracted by 100,000 volts would fly through the aether at a speed of 164 million m/s, and its particle-wave duality frequency would be in the X-ray range ($4.44 \times 10^{19}$ Hz). In other words, the EPs are shoved aside by the electron, and they close ranks again behind the electron; this is associated with a shock front that vibrates at a frequency of $4.44 \times 10^{19}$ Hz. No changes in spin are involved; in accordance with Fig. 9-1, the spin of the EPs in this case would correspond to the 100,000-volt electric field.

A missile in the air gives up energy as it generates a thermal shock wave. But an electron doesn't give up energy as it shoves EPs aside, and the X-ray shock wave carries no energy; it cannot expose a photographic film. We know that it exists from single-electron diffraction and interference effects [24].

# CHAPTER 10

▼

# HOW THE AETHER WAS
# REPEATEDLY ABANDONED

From my perspective, here is what the universe looks like: it is occupied by aether particles, which define the universe. Here and there in this huge ocean of particles, like grains of sand, are minuscule atomic nuclei, mostly protons. At relatively large distances, like planets around the sun, we have electrons orbiting the nuclei. So the universe consists of "aether" space with small interlopers, material objects such as stars, planets, people, atoms, and electrons. These move about at a relatively slow pace compared to the aether's natural speed of $3 \times 10^8$ m/s. For example, the movement of the Earth around the sun is at a speed 0.01% of the speed of light. As it moves, each material object causes a slight turbulence in the ocean of aether particles.

Some of the above is depicted in Fig. 1-1, which shows the Earth (*US*) and a second planet that is moving to the right (*THEM*). In accordance with Einstein's special relativity, the velocity of light on *each* planet is $3 \times 10^8$ m/s; this is accomplished by having an aether "atmosphere" surround each planet, held in place by gravity exactly as our air atmosphere (and "dark matter"?) is held in place. Between the planets is the aether background, slowly drifting, say, to the north.

Because the aether is a necessity and not a conjecture, heroic efforts were required, starting in 1905, to justify the abandonment of the aether. It is especially interesting to see how the textbooks, charged with conveying the truth to their inquisitive and sometimes skeptical audiences, weighed "the facts in the case." Not a single one of them suggested that, in the strange and elusive world of the aether particle, it was possible for an electromagnetic field to propagate across an aether motion discontinuity in a straight line, without bending, as in Fig. 1-1(b), and that this is why it is so difficult to detect the aether.

What was it like to be alive in 1905? Peter Galison's entire book, *Einstein's Clocks, Poincaré's Maps*, describes this fascinating period [15]. Albert Einstein (1879–1955) was twenty-six. Henri Poincaré (1854–1912) was fifty-one. Hendrik A. Lorentz (1853–1928) was fifty-two. In Galison's index, there are twenty-six page references to "Aether—Einstein's rejection of." The same thought repeated twnety-six times can get very boring, but Galison is too skilled a writer to let that happen. In the caption that describes the Michelson-Morley (M-M) apparatus, Galison neatly sums it up with the following paragraph (p. 204):

**Hunting the Aether**. With a remarkable series of experiments, Albert Michelson sought to measure the earth's motion through the elusive aether. In the 1881 device shown here, he launched a beam of light from *a* that was split by a half-silvered mirror at *b*: one-half of the ray reflected off *d* and into the eyepiece *e*. The other half of the ray penetrated the mirror at *b*, reflected from *c*, and was then bounced from *b* to the eyepiece *e*. At the eyepiece the two rays interfered with each other, showing the observer a characteristic pattern of light and dark. If one wave was delayed—by so little as a part of a wavelength of light—this pattern would visibly shift. So if the earth really was flying through the aether, then the "aether wind" would affect the relative time it took for the two beams to make their round-trips (the relative phase of the two waves would shift). Consequently, Michelson fully expected that if he rotated the apparatus, he would see a change in the interference patterns of the two rays. But no matter how he twisted his staggeringly sensitive instrument, the dark and light patterns did not budge. To Lorentz and Poincaré, this meant that the interferometer arms—like all matter—were contracted by their rush through the aether in just such a way as to hide the effect of the aether. To Einstein it was one more suggestive piece of evidence that the very idea of the aether was "superfluous."

Next, jump ahead fifty years, to 1955, when Wolfgang K. H. Panofsky and Melba Phillips very thoroughly examined aether theories in their book, *Classical Electricity and Magnetism* [19]. (A human-interest note: Melba Phillips died on November 8, 2004, at the age of 97.) On page 240 is a table that compares three aether theories and special relativity (without an aether) to thirteen experimental outcomes (including the Michelson-Morley experiments). One of the aether the-

ories is called "aether attached to ponderable bodies," which is the theory pro-
moted in this book. On pages 236–237, Panofsky and Philips state the following:

> **"Aether drag."** A further alternative in which the concept of the aether
> could be reconciled with the Michelson-Morley result was to consider
> the aether frame attached to ponderable bodies. This would automati-
> cally give a null result for terrestrial interferometer experiments. The
> assumption of a local aether, however, is in direct contradiction to two
> well-established phenomena.
>
> The first is the aberration of "fixed" stars. Due to the motion of the
> earth about the sun, distant stars appear to move in orbits approximately
> 41" in angular diameter. Consider a star at the zenith of the ecliptic. If
> this star is to be observed through a telescope the telescope tube must be
> tilted toward the direction of the earth's motion by an angle $\alpha$, as shown
> in…. It is seen from the figure that, classically, $\tan \alpha = v/c$, and with 30
> km/sec for the velocity of the earth in its orbit, $\alpha = 10^{-4} = 20.5"$, in
> agreement with observation. If the aether were dragged by the earth in its
> motion we should expect no aberration to occur.

The model of Fig. 1-1(a) fails the stellar aberration test, which I claim is
explained by the straight-line photon locus of Fig. 1-1(b). The model of Fig. 1-
1(b) also explains the second phenomenon mentioned above, which is too technical to repeat here. Panofsky and Phillips conclude that "[t]he existence of an
aether, either stationary or carried convectively, is undemonstrable."

Consider next an entire book devoted to Einstein's special relativity: N. David
Mermin's *Space and Time in Special Relativity* [11]. On page 13 of this book,
Mermin states the following:

> In essence, the famous Michelson-Morley experiment was an attempt to
> measure [the] directional dependence of the speed of light with respect
> to the Earth and thus to determine the speed of the Earth with respect to
> the aether. The result of their experiment was that the speed of light with
> respect to the Earth has the same value $c$ whatever the direction of
> motion of the light.
>
> One might try to explain this by saying that the speed of the Earth
> with respect to the aether must be zero. Aside from the fact that this
> would be a rather strange coincidence, this explanation will not do. The
> Earth moves in its orbit around the Sun at about 30 kilometers per sec-
> ond. If the velocity of the Earth with respect to the aether happened to
> be zero at one time of year, then 6 months later when the Earth was mov-
> ing at the same speed but in the opposite direction, its speed with respect

to the aether would have to be 60 kilometers per second. In general, because of the Earth's motion around the Sun, whatever the speed of the Earth with respect to the aether might be, this speed should vary through a range of speeds differing by up to 60 kilometers per second, throughout the course of a year. However experiments have shown that the speed of light with respect to the Earth is independent of direction, whatever the time of year.

Thus if the aether does exist, it must be managing in a most mysterious way to escape our efforts to detect it. As Einstein showed, the way out of this dilemma is to deny the existence of the aether and face courageously the fact that light moves with a speed $c$ with respect to any inertial observer whatsoever, regardless of the velocity of that observer with respect to any other observer (and, as a special case of this, regardless of the velocity of the source emitting the light).

My comment is that the aether moves *with* the Earth, and with every massive (ponderable) object, much as the Earth's air atmosphere moves with the Earth. And it does not take any courage to believe that the speed of light on a planet (or any massive object) moving away from (or toward) the Earth is $3 \times 10^8$ m/s *relative to that planet*, such as the planet shown in Fig. 1-1.

Closer to our own time, let's look at the Robert Mills's book, *Space, Time and Quanta*, which was published in 1994 [12]. On page 17, Mills states the following:

The big issue at the time [1905] was the motion of the earth relative to the *aether*, the supposed material substance thought to fill all of space and to act as a medium for the propagation of light, analogous to the role of air with sound waves. Experiments such as Michelson and Morley's were known as aether-drift experiments and were interpreted, though with severe difficulties, as indicating that the aether is dragged along with the earth. The concept of an aether was buried by Einstein (it was already pretty sick by then), and I shall not discuss its history any further here. As we now understand, EM [electromagnetic] waves require no medium, and they exist and propagate even in completely empty space.

Next, we expect the *McGraw-Hill Encyclopedia of Physics* to reveal the life, and perhaps death, struggle of the aether [25]. Alas, it is not a history book. On page 392, William R. Smythe wrote three short paragraphs on the subject:

### Aether hypothesis

James Clerk Maxwell and his contemporaries in the nineteenth century found it inconceivable that a wave motion should propagate in empty space. They therefore postulated a medium, which they called the aether, that filled all space and transmitted electromagnetic vibrations.

During the last half of the nineteenth century, dozens of models were tried, but all broke down at some point. Direct experimental attempts to establish the existence of an absolute aether frame of reference, in which Maxwell's equations hold and light has the velocity $c$, have failed. The best known of these is the Michelson-Morley experiment, in which an attempt was made to measure the velocity of the Earth relative to the aether.

Every hypothesis (aether drag, Lorentz contraction, and so on) invented to reconcile some experiment with the aether concept has been disproved by some other experiment. At present, there is no evidence whatever that the aether exists.

My comment is that Einstein's special relativity, in which the velocity of light is $c$ regardless of the speed with which a planet is flying away (or toward) the Earth, is the physical evidence that the planet carries an "atmosphere" of aether, as shown in Fig. 1-1.

As a final example of the abandonment of the aether, *Science* observed its 125th anniversary with a section titled "125 Questions: What Don't We Know?" (volume 309, p. 75, 1 July 2005). In the introductory essay by Tom Siegfried, "In Praise of Hard Questions," he writes, "When *Science* turned 20…Maxwell's mentor William Thomson (Lord Kelvin) articulated the two grand gaps in knowledge of the day….One was the mystery of specific heats that Maxwell had identified; the other was the failure to detect the aether, a medium seemingly required by Maxwell's electromagnetic waves." The aether is not one of the 125 questions, but this is standard treatment for a concept that was abandoned 100 years ago. Let's hope that the aether is "discovered" before *Science's* "150 Questions" issue.

# CHAPTER 11

# THE DOUBLE-SLIT PHOTON INTERFERENCE PATTERN

As a strategy for studying the photon, we start out, innocently enough, with the relatively strong electromagnetic field (EMF) output of a laser, and then attenuate the field until individual photons can be isolated. Truly strange and unbelievable happenings are then observed [22].

As a vehicle for this discussion, consider the double-slit (sometimes called the two-slit) diffraction-interference apparatus shown in Fig. 11-1(a). The EMF, polarized with the $\underline{E}$ lines in the plane of the page, as shown, is propagating to the right. It strikes an opaque plate that contains two slits. (They are at right angles to the page). Some of the EMF gets through the upper slit, and some gets through the lower slit.

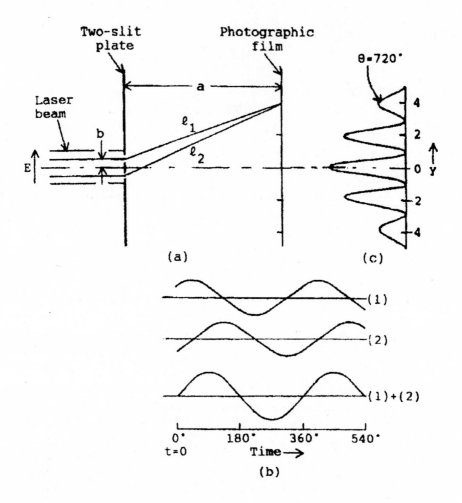

**Fig. 11-1.** Double-slit interference and diffraction. (a) This shows a schematic of the apparatus. The slits are at right angles to the page. Two of the rays leaving the slits are depicted as they meet at $y = 4$ of the photographic film. (b) This shows the waveforms of rays (1) and (2) when they meet at the film if they are 90° out of phase. (c) This shows the film pattern. The *same* pattern results if the input "beam" consists of single, one-at-a-time photons.

The effects exploited here are used in many different applications; an especially fruitful area is that of astronomy. The two words, diffraction and interference, may seem ominous, but the ideas are really very simple. We are dealing here with sine waves, such as the $\underline{E}$ and $\underline{H}$ waveforms shown in Fig. 4-1. When the laser beam sine wave of Fig. 11-1(a) passes through a narrow slit, it spreads out laterally—it diffracts—so that light passing through each slit spreads over the photographic film at the right. The film intercepts light extending from $y = -5$ to $y = +5$, as defined by the vertical scale of Fig. 11-1(c).

The second word, interference, is misleading, but it is too late for us to change it. "Interference" implies that the two rays emerging from the slits act to hinder or impede each other. This is fine for football, but in Fig. 11-1(a), half of the time, the two rays aid each other (constructive interference). This seems to be a good example of an oxymoron. The process in which the two rays hinder each other is called "destructive interference."

Two of the rays thus formed, (1) of length $\ell_1$ and (2) of length $\ell_2$, are singled out as they come together on the sheet of photographic film. (Visible or ultraviolet light is usually used because their photons have sufficient energy to be recorded on the film.) What pattern will the exposed film show?

In some locations, the EMF from ray (1) is in phase with that of ray (2) when they meet at the film, thus increasing film exposure (constructive interference). At other locations, they have opposite phases, and the EMFs cancel (destructive interference). Fig. 11-1(b) illustrates an in-between situation in which they are 90° out of phase; there is some increase in the total output; the output increases by a factor of 1.414. The net result of constructive and destructive interference can be seen in the set of peaks and valleys shown in Fig. 11-1(c).

Rays $\ell_1$ and $\ell_2$ are shown with the following relative values (as defined in Fig. 11-1): $b = 0.5$, $a = 10$, $y = 4$, and $\theta = 720° = 4\pi$ (because it is the second peak away from the $y = 0$ axis). The numerical values correspond to a relative laser light wavelength $\lambda = 0.1857$. Ray $\ell_2$ is 10.97 units long and contains 59 cycles of laser signal. Ray $\ell_1$ is 10.59 units long and contains 57 cycles. Therefore, the two signals arrive in phase (constructive interference).

At $y = 1$ ($\theta = 180° = \pi$), the longer path is 54.4 cycles long and the shorter path is 53.9 cycles long, so the difference is 0.5 cycle. Therefore, the two signals cancel (destructive interference).

The film pattern follows a commonly encountered intensity pattern [(sin $y$)/($y$)]$^2$; that is, it is a sine squared wave (always positive) whose amplitude decreases as $y$ increases.

Now consider that the EMF is a form of energy. Where the EMF vanishes because of destructive interference, its energy must be picked up by regions of constructive interference. From a photon's point of view, a photon is a form of energy ($E = hf$). It travels at right angles to its $\underline{E}$ (and $\underline{H}$) lines. After each photon gets past the double slits, it diffracts by an amount that is based on its predetermined but statistically random prior experiences. Because of the bending of the $\underline{E}$ lines, entering photons veer off toward the upper and lower paths, avoiding the middle destructive-interference path. When a photon strikes the photographic film, its energy is released, exposing a small dot (diameter approximately equal to the photon's wavelength).

In other words, in Fig. 11-1(a), the photons actually curve away from destructive-interference points $y = \pm1$ and $y = \pm3$, and toward constructive-interference points $y = 0$, $y = \pm2$, and $y = \pm4$. As a result of this "curving away," the valleys of Fig. 11-1(c) are created.

### Simultaneous-Burst Pattern

Our next step is to carefully decrease the output of the laser beam. Suppose that an ideally fast shutter allows a burst of only 1000 photons to *simultaneously* fly through the slits. We are immediately faced with probabilities. Around 500 photons will probably pass through the upper slit, and the remaining approximately 500 photons will pass through the lower slit. Their $\underline{E}$ and $\underline{H}$ fields join up, + to −, as they laterally disperse via diffraction.

Experiments show that the film exposure display of Fig. 11-1(c) occurs independent of laser beam intensity (but not much, if anything, will be visible if there is a total of only 1000 photons).

In order to refer to specific numerical values, the outline of a reasonable distribution diagram is shown in Fig. 11-2(a). If the bins are $0.5y$ unit wide, the summation of all the values ($16 + 21 + 28 +...$) equals 1000 (except for a rounding-off discrepancy). The result is a crude approximation, but it is adequate for my purpose. Out of the 1000 photons, 75 will head for the $y = 0$ bin, 74 for each of the $y = \pm0.5$ bins, 71 for each of the $y = \pm1$ bins, and so forth. These are reasonable values, and would in fact appear as a film exposure if only a single slit were open and the interference mechanism could not operate.

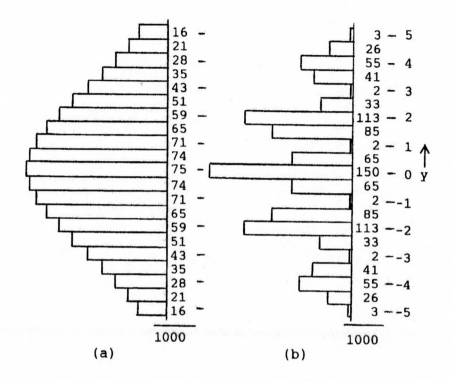

**Fig. 11-2.** Photon exposure distributions at the film of Fig. 11-1(a) if the bins are $0.5y$ unit wide. (a) This shows the photon exposure distribution due to an assumed diffraction attenuation function, $\exp(-0.0625y^2)$, with the interference effects omitted. (b) This shows the photon exposure distribution including constructive and destructive interference, as shown in Fig. 11-1(c).

The procedure used to derive Fig. 11-2(a) was applied to the film exposure display of Fig. 11-1(c), yielding Fig. 11-2(b). Here, out of the 1000 photons, 150 end up in the $y = 0$ bin. This is reasonable if half of the $y = 0.5$ and $y = -0.5$ photons, from Fig. 11-2(a), are captured by the $y = 0$ bin. But what happens to the 71 photons that, according to Fig. 11-2(a), started out headed for $y = 1$? Fig. 11-2(b) tells us that only 2 photons get through. What happens to the other 69 photons? They end up in the constructive-interference regions to either side of $y = 1$.

### Individual-Photon Pattern

Finally, instead of 1000 simultaneous photons, we block the light so effectively that *only one isolated photon at a time gets through*—one per second, say. After 1000 seconds (16 2/3 minutes), we develop the film. We expect to see Fig. 11-2(a) because constructive or destructive interference could not possibly occur with individual one-at-a-time photons. Instead, however, *we get Fig. 11-2(b)!*

This is an unbelievable result, impossible to explain by classical physics. It defies common sense.

We are fortunate to have a computer simulation of the actual diffraction pattern, in Fig. 11-3, for a total of 10,000 individual photons. This is the work of Tore Wessel-Berg, found in his book, *Electromagnetic and Quantum Measurements* [26] (p. 185). In Wessel-Berg's book, some of the details surrounding the computer simulation are given along with the figure. Before you object that "this is only a computer simulation," I can assure you that it is exactly like the photos reproduced by Tonomura et al. [24]. Tonomura's photos are for electron diffraction (see Chapter 15), but there is no difference between electron and photon diffraction where the film output is concerned. The computer simulation generates a high-contrast printout, and avoids the tremendous experimental difficulties that Tonomura et al. had to overcome.

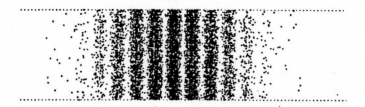

**Fig. 11-3.** Computer simulation, based on the bitemporal theory of Professor Wessel-Berg, of the build-up of a diffraction pattern by 10,000 one-at-a-time photons dumping their energy, $E = hf$, at localized spots on the screen. This figure is copied from Tore Wessel-Berg's *Electromagnetic and Quantum Measurements: A Bitemporal Neoclassical Theory* with the kind permission of Springer Science and Business Media [26] (p. 185).

The evidence would have us believe that each photon somehow divides in half, and each half goes through a slit. Upon emerging from the slit, each half is apparently associated with an EMF that is similar to that of 1000 simultaneous photons (except, of course, that the total EMF energy is that of a single photon). The emerging EMFs cover the entire film of Fig. 11-1(a), from $y = -5$ to $y = +5$. The energy of the EMF that strikes the film should be modified by constructive and destructive interference, as depicted in Fig. 11-1(c). Instead, the photon behaves like a point particle, lands on the film at $y = 4$ (for example), and *all* of its energy is converted into a single bright dot at $y = 4$. After 1000 seconds, it will turn out that some 55 photons [a value given by Fig. 11-2(b)] were captured by the $y = 4$ bin; 150 landed in the $y = 0$ bin; and so forth.

There are two serious problems with the above recital. First, since a photon is the "irreducible constituent" of an EMF, it cannot split into two halves, each passing through one of the slits. Second, if the photon gives birth to an EMF that covers the entire film from $y = -5$ to $y = +5$, the photon's energy would reside in this field, leaving much less than a normal amount for the wave packet that eventually strikes and exposes the film at $y = 4$.

These problems have confounded physicists for many years. Much of Nick Herbert's *Quantum Reality* [27], Jim Baggott's *The Meaning of Quantum Theory* [28], David Lindley's *Where Does the Weirdness Go?* [29], and Robert Mills's *Space, Time and Quanta* [12] are devoted to various explanations, with various degrees of plausibility. The difficulty is that there is no satisfactory realistic theory, as I have stressed above, based on quantum mechanics or classical physics. Quantum mechanics is inappropriate for describing the behavior of an individual photon or electron; it is superb for revealing the statistics for many photons or electrons. One must conjecture outside the limits of classical or quantum physics.

# CHAPTER 12

▼

# THE WAVE-PARTICLE DUALITY FIELD

In what follows, the existence of a field that is analogous to an electromagnetic field is proposed; it is called a wave-particle duality field, or WPD field. My conjecture is that it is a type of compression shock wave generated as the photon plows through the aether. (Although it is nominally a "compression" wave, it actually consists of compressions and expansions). This is analogous to a high-speed projectile traveling through the air. Air supports the propagation of sound waves, and a projectile forms a shock wave. The shock wave consists of compressions (and expansions) propagating at the speed of sound. Constructive and destructive interference always show up when the shock wave reaches a reflecting object or refractive medium.

Analogously, the aether supports the propagation of EMFs, and the photon "projectile" forms a shock wave that propagates at the speed of the EMF. It would be premature, however, to think that the WPD field really is a shock wave. We know a great deal about air and sound shock waves, but we do not know what the aether, electric field, and magnetic field really are. Despite this ignorance, we get through life drawing electric and magnetic field lines and designing sophisticated equipment based upon imaginary field intensities and flux densities. In the world of imaginary field lines that follows, we assume that WPD field lines really exist because they are associated with experimentally revealed constructive and destructive interference.

However, one should not pursue the analogies too far. A sonic boom carries a tremendous amount of energy, but the WPD field may not carry any energy at all. Zero energy? The aether is a peculiar medium: we peer at photons, tiny wave-packets that have been traveling for billions of years through the aether *with zero attenuation*. From another viewpoint, there can be no attenuation because the latter implies the conversion of photon energy into heat, which in turn implies that some particle that has mass (such as an atom) will vibrate more rapidly as it absorbs this energy. But there are no atoms in the aether, or at least none that have absorbed the energy of this billion-year-old wave packet (which is why we can detect it, of course). In other words, the aether is a perfectly elastic, lossless, linear medium; the transverse ripple of Fig. 4-2 is passed along, without change, at the velocity of propagation.

Closer to home, and something about which we know a great deal, there is the zero attenuation for superconductivity and superfluidity.

For many electrical conductors (and, recently, semiconductors), if they are cooled toward 0 K, a transition temperature is reached at which electrical resistance vanishes. Other changes also take place at the transition temperature: magnetic fields are expelled, and thermal properties are altered. The theoretical explanation for superconductivity was presented, in 1957, by J. Bardeen, L. N. Cooper, and J. R. Schrieffer.

Helium liquefies at 4.22 K. If it is further cooled, to 2.172 K, a transition occurs at which viscosity vanishes. The superfluid is able to flow at high speed through tiny holes. Here, also, other changes take place at the transition temperature.

Before the days of superconductivity and superfluidity, we could not conceive of zero electrical resistance and zero viscosity. They were amazing experimental discoveries. (Superconductivity was discovered by H. K. Onnes in 1911.) In this same spirit of open-mindedness, we may conjecture that the WPD field can certainly be a zero-energy field if it is not required to do work. From here on, in this book, it is conjectured that the WPD field shock wave consists of compressions (and expansions) of the aether that do not convey any energy.

Is the WPD field a transverse vibration, like the wave packet in Fig. 4-2, or is it a longitudinal vibration, which we expect for a "compression shock wave," as shown in Fig. 3-1? There are arguments that support either conjecture. A longitudinal vibration is illustrated at the end of this chapter.

A photon has a dual personality. On the one hand, it is the wave packet of Fig. 4-2; the wave represents transverse electric and magnetic propagating fields. The direction of the electric field is, by definition, the direction of polarization. Energy resides in the fields: $E = hf$. Electric and magnetic fields are forms of energy. Apply a voltage to two parallel conducting plates, and a current flows to charge the plates and create an electric field between them. Some of the energy in the current is thus converted into electric field energy. Similarly, apply a voltage to a coil of wire, and a current flows to create a magnetic field in and around the coil. This time, some of the energy in the current is converted into magnetic field energy.

The other half of the photon's "personality" is that it behaves like a *particle* that is hurtling through the aether at $3 \times 10^8$ m/s. Visualize the aether particles as being shoved aside to make way for the photon projectile. Then, as the aether particles close in behind the projectile, the WPD field oscillates at the same frequency as that of the photon. But this could not possibly be correct because, when the power pack emerges from one of the double slits, it is guided toward a constructive interference point. Therefore, the WPD field must lead the way; it must be a compression (and expansion) zero-energy WPD field that forms *in front* of the power pack.

How far does the WPD field extend in front of the power pack? The WPD field must extend at least 10 or 20 wavelengths in front of the power pack, enough to get a reasonably effective degree of destructive interference. The WPD field may therefore be finite, like the strong-force field of an atomic nucleus.

Views (a) through (g) of Fig. 12-1 depict how the photon WPD model can explain the single, isolated photon double-slit experimental results of Fig. 11-1. (The slits are greatly magnified for the sake of clarity.)

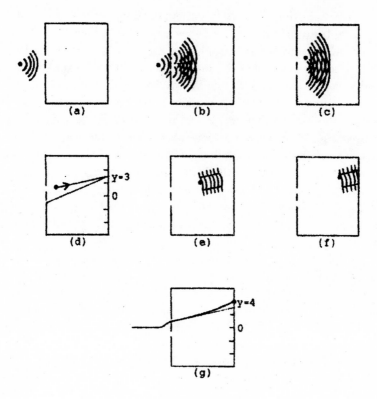

**Fig. 12-1.** Sequence that illustrates double-slit interference effects that accompany a single, isolated photon. (a) This shows a photon approaching the slit plate. (b) This shows how the leading portion of the WPD field has split, with a fragment getting through each of the slits. (c) This shows how the WPD fields have progressed beyond the slit plate. The power pack, because of a predetermined but statistically random past history, has followed the upper-slit WPD segment. (d) This shows the same situation shown in (c), but with WPD fields omitted. The power pack is heading for the $y = 3$ point of the photographic film. (e) This shows the power pack and net WPD field, halfway across. (f) Because WPD field lines are concave, the power pack is directed away from the destructive-interference $y = 3$ point. (g) This shows the power pack locus curves, exposing the film at the $y = 4$ point. The ethereal WPD field has vanished without a trace.

In Fig. 12-1(a), the photon is approaching the double-slit apparatus.

In Fig. 12-1(b), the leading portion of the WPD field has split, with a fragment getting through each of the slits. The fragments diffract. Thus far, the action is identical to that of a laser beam directed at the two slits.

In Fig. 12-1(c), we depart from a conventional perspective. The power pack and at least some of the WPD field are inseparable, since it is impossible to generate a shock wave without the power pack. In getting to Fig. 12-1(c), the photon has three choices: (1) the photon can strike the slit plate at the center, in which event the photon's energy is converted into heat and the WPD field vanishes without a trace; or (2) the photon can pass through the upper slit [the choice shown in Fig. 12-1(c)]; or (3) the photon can pass through the lower slit. The actual path taken by the photon is predetermined but statistically random, based on its prior history.

There is a serious problem here with regard to the lateral movement from Fig. 12-1(b) to Fig. 12-1(c). Because the photon has zero mass, one may think it can be pushed sideways without the expenditure of force. This is not so for a photon that, after all, travels at the speed of light. The *effective* mass is determined by $E = mc^2$, which, combined with Planck's law, gives the effective mass:

$$m_{eff} = hf/c^2. \quad (12\text{-}1)$$

For the photon generated when an electron spirals from the $n = 2$ to $n = 1$ orbit of a hydrogen atom, $f = 2.467 \times 10^{15}$ Hz, so Equation 12-1 yields $m_{eff} = 1.819 \times 10^{-35}$ kilogram.

This is a truly minuscule mass. It is 50,000 times lighter than an electron. (This should be kept in mind by those who are designing equipment to detect through which of the two slits the photon traveled. An effective photon mass that equals that of an electron is obtained with a frequency of $1.236 \times 10^{20}$ Hz; this is on the borderline between X-rays and gamma rays.)

Nevertheless, despite its minuscule effective mass, a finite force has to act on the photon to achieve lateral deflection. If the double-slit experiment is performed using a laser beam, there is plenty of energy in the EMF to support lateral movement, but not with a single, isolated photon.

Although it may not be valid to think of the photon as being similar to a high-speed projectile in air, the analogy suggests a solution to the lateral-force problem. The conjecture is that the aether forms streamlines through the two slits, and these guide or steer the photon. The aether supplies the lateral force, much as a glancing blow can force a projectile in air to change its course. There is no change in kinetic energy if no change in speed is involved, so the lateral push need not entail a change in energy.

The lateral force is reminiscent of the force of attraction between two conducting, uncharged plates brought sufficiently close together in a high vacuum.

The minuscule force is known as the H. B. G. Casimir effect. It may be possible that this force, which has been measured by S. K. Lamoreaux [30], is another zero-energy phenomenon.

## Streamlines

What are streamlines? In smoothly flowing water (a nonturbulent "stream"), they trace out the flow lines. Think of the aether as flowing through the slits. This implies that the aether is not a passive jelly. The conjecture here is that the aether is a perfectly elastic medium in which streamlines are ubiquitous. The streamlines in an all-pervading aether guide the compression shock waves; this is reminiscent of the pilot wave proposal made by David Bohm [31].

Returning to Fig. 12-1, Fig. 12-1(d) is the same as Fig. 12-1(c), except that the WPD fields are omitted for the sake of clarity. We now see that the particular WPD field fragment to which the power pack was attached, in Fig. 12-1(c), has directed the power pack to $y = 3$.

In Fig. 12-1(e), the power pack is midway between the double-slit plate and the photographic film. Because it is approaching a destructive-interference point, the WPD field lines are concave. This translates into aether streamlines that laterally push or "encourage" the power pack to head for the constructive-interference points at $y = 2$ or $y = 4$.

In Fig. 12-1(f), the power pack is shown on a path toward $y = 4$.

In Fig. 12-1(g), the power pack arrives at the film, exposing a tiny dot at the $y = 4$ position. According to Fig. 11-2(b), if 1000 individual photons are launched in this way, in sequence, 55 of them will end up in the $y = 4$ slot, and only 2 will end up in the $y = 3$ slot.

Fig. 12-1(g) shows the path taken by the power pack. The various curves are explained by the lateral forces exerted by the aether upon the photon. The WPD field is an ethereal compression shock wave; it vanishes without a trace.

If the aether streamlines are moving to the right at the speed of light, what happens when they strike the photographic film? Nothing much at all. Remember that a material object, like the film, is mostly empty space with a few atomic nuclei here and there. The photon's power pack, on the other hand, is stopped by these nuclei, and transfers its kinetic energy to them. Perhaps a tiny probe, to the right of the film, would be able to pick up the aether particle wind as it emerges from the film.

It is also possible that the streamlines are almost stationary, and exert a lateral force without appreciably moving.

But one important loose end remains: that of the change in momentum (the mass multiplied by the change in velocity). If the power pack changes course in order to land at $y = 4$ in Fig. 12-1(g), the change in momentum must be balanced by an equal and opposite change in the momentum of the ethereal streamlines. This concept is greatly assisted if an aether particle has mass and inertia. (Perhaps it is that elusive dark matter, after all.) We realize, from the above, how much more there is to discover about the aether.

The concept that the WPD field may be a longitudinal wave is depicted in Fig. 12-2(a). To the right of the power packs, black and white strips symbolize the compression and expansion of the aether, respectively. The split paths, in which the power pack proceeds through the upper slit, is illustrated.

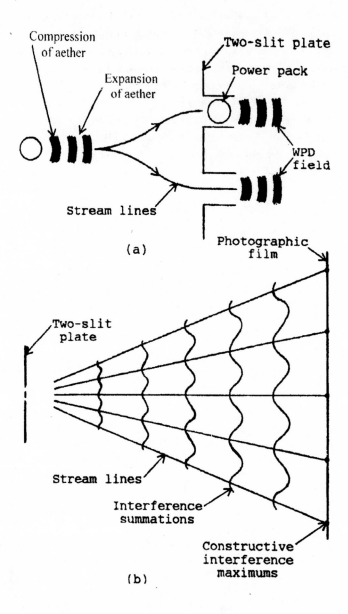

**Fig. 12-2.** Additional illustration of double-slit plate interference. (a) The WPD field is depicted as a longitudinal wave. (b) The ethereal streamlines follow the interference maximum summation peaks.

In Fig. 12-2(b), the WPD lines interfere; the ethereal streamlines follow the interference maximum summation peaks. These correspond to regions where the $\underline{E}$ field intensity is maximum; it is these points that guide the streamlines, which "encourage" the power packs to end up near constructive interference maximum points.

## Comment

What do physicists have to say about the single-photon results of Fig. 11-1? With the kind permission of Springer Science and Business Media, I would like to quote a physicist (to contrast with an engineer). In physicist Tore Wessel-Berg's *Electromagnetic and Quantum Measurements: A Bitemporal Neoclassical Theory*, he states the following [26] (pp. 151–152):

> The problem arises when the intensity of the incident radiation is gradually reduced until only one single photon at a time arrives at the slit plate. Such a condition can be realized experimentally and is not a theoretical construct. Under these circumstances it is observed that the well separated individual photons dump their energy at separate localized spots on the photosensitive screen. This observation, quoted in any textbook on quantum physics, certainly projects the notion of a particlelike behavior of the photon. On the other hand, as more and more photons are arriving at the screen, the accumulated set of spots builds up to the typical diffraction pattern.... And this experimental result, being in full agreement with optical theory based on wave propagation, certainly points to the photon as a wave. These contradictory statements—the photon behaving as both a particle and a wave—are the source of the celebrated dilemma in quantum physics, and a major contribution to the launching of the duality principle. This principle interprets the photon as a dual particle, behaving sometimes as a wave and sometimes as a particle, depending on circumstances. The argument goes something like this: the photon has to possess wave nature in order to pass through both slits, which is necessary in order to produce the observed diffraction pattern. On the other hand, the experiment shows that each photon arrives at the screen as a localized pointlike entity producing single and well separated black spots on the screen, thereby indicating a particlelike behavior. If the photon is a particle, it must have gone through one of the two slits, in clear contradiction to the first statement. This is the celebrated "which way" paradox. Does the photon, now as a particle, go through the first or the second slit? And this problem is accentuated with the additional paradox that the particle really has to go through both slits in

order to produce the diffraction pattern. This paradox is a major dilemma in quantum physics, and lots of ink has flowed in futile attempts to resolve the paradox.

My comment is that Professor Wessel-Berg *does* have a very complete explanation for the "which way" and "diffraction pattern" quantum paradoxes; that is, his "bitemporal hypothesis." Note that the subtitle of his book is "A Bitemporal Neoclassical Theory."

# CHAPTER 13

▼

# AN INTERFEROMETER EXPERIMENT

This chapter continues with the discussion of an experiment that yields a result that cannot be explained by any existing reality, but which can be explained by the WPD concept. This experiment is referred to in Fig. 13-1, which is discussed by Paul Kwiat et al. [32]. Fig. 13-1 depicts a "thought experiment" suggested by Avshalom C. Elitzur and Lev Vaidman, but Kwiat and his colleagues have verified the concept in a relatively complicated laboratory setup. Only the much simpler thought experiment will be considered.

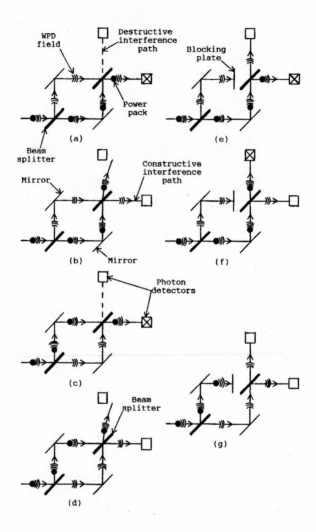

**Fig. 13-1.** Interferometer experiment that yields a strange result [32], but that can be explained by the WPD model. In each part [(a) through (g)], a single, isolated photon enters at the lower left corner. (Although four photons are usually shown, they are the *same* photon "photographed" at different stages of its flight.) The photon is processed by two beam splitters, two mirrors, and two photon detectors. A photon striking a detector is symbolized by an ×. The entering photon has a 25% probability of following each of the scenarios, (a) through (d), in the left column. If a blocking plate is added as shown, (a) becomes (e); (b) becomes (f); and (c) and (d) become (g). The strange result is demonstrated by (f): although the blocking plate does not intercept any photon energy, the photon is "seen" because the upper detector registers an ×.

In each of the seven parts [(a) through (g)] of Fig. 13-1, a *single*, isolated photon enters at the lower left corner and strikes a beam splitter. The latter is analogous to an imperfect mirror: about half of the photons that strike the beam splitter will pass through to the right, as in Fig. 13-1(a) and Fig. 13-1(b); the other half are subjected to a mirror-type reflection, as in Fig. 13-1(c) and Fig. 13-1(d). (One can identify the photon by its "power pack," of course. Although four photons are shown in almost every part, they are the *same* photon "photographed" at different stages of its flight.)

The apparatus of Fig. 13-1 contains a second beam splitter. In Fig. 13-1(a), the photon, moving upward, strikes the second beam splitter and is reflected to the right. In Fig. 13-1(b), however, it passes through and continues to move in an upward direction. In Fig. 13-1(c), the photon, moving to the right, strikes the second beam splitter and passes through, continuing to move to the right. In Fig. 13-1(d), however, it is reflected in an upward direction.

The entering photon has a 25% probability of following each of the four parts [(a) through (d)] in the left column of Fig. 13-1.

The experiment requires two photon detectors, as shown. A photon striking a detector is symbolized by an × in the detector box.

I will now ask you, the reader, to add your own "thought experiment" to the thought experiment. Please erase any paths except those containing three arcs with a power pack (the photon). This is the spirit in which the article by Kwiat et al. is written. However, they imply the presence of wave-particle duality fields (the three arcs without a power pack), without actually admitting that WPD fields exist, because the apparatus is an interferometer. It demonstrates constructive and destructive interference. (You will recognize that it is a variation on the theme represented by the double-slit plate of Fig. 11-1.)

A key element in Fig. 13-1, however, is that the right-hand photon detector is on a constructive interference path, while the upper photon detector is on a destructive interference path, as shown. As Kwiat et al. put it, the "Elitzur-Vaidman experiment gives a photon a choice of two paths to follow. The optical elements are arranged so that photons always go to detector D-light (corresponding to constructive interference) but never to D-dark (corresponding to destructive interference)."

Next, let us conjecture that the photon is accompanied by a WPD field. Then the outcome will make sense, and parts (a) through (d) of Fig. 13-1 can be described as follows below.

In Fig. 13-1(a), when the entering photon strikes the first beam splitter, it continues to move to the right, preceded by its WPD field. A remnant of the WPD field is reflected upward and then to the right. The power pack's WPD field, and the "remnant" WPD field, meet at the second beam splitter. They are in phase at the right-hand detector, generating an ×. But now a second remnant of the power pack's WPD field encounters the first remnant on the path to the upper detector. These WPD fields are 180° out of phase, and cancel each other.

In Fig. 13-1, 25% of the entering photons will follow the locus depicted in Fig. 13-1(b). First and second remnants of the power pack's WPD field are in phase and reach the right-hand detector; because they are zero-energy compression shock waves "generated as the photon plows through the aether," they vanish without a trace. At the path to the upper detector, however, a minor complication shows up: because the WPD fields cancel, the power pack, which cannot turn back, veers off to the right (or left) along a *constructive* interference path. Recall that the power pack represents energy, $E = hf$, that cannot simply vanish like a WPD field. It is conjectured that ethereal streamlines guide the power pack to the right (or left) to avoid the destructive-interference (dashed) path.

In Fig. 13-1(c) and Fig. 13-1(d), the actions at the right-hand and upper photon detectors are a repeat of those in Fig. 13-1(a) and Fig. 13-1(b), respectively.

Now, here comes the important and interesting change: a blocking plate is added as shown, interrupting the upper path between a mirror and the second beam splitter. (Kwiat et al. used an exploding pebble rather than a plate, perhaps to add excitement to a recitation that may otherwise be dull, but I am less imaginative.) In Fig. 13-1, with the plate, Fig. 13-1(a) becomes Fig. 13-1(e); Fig. 13-1(b) becomes Fig. 13-1(f); and Fig. 13-1(c) and Fig. 13-1(d) both become Fig. 13-1(g). This is discussed below.

In Fig. 13-1(e), the remnant of the entering photon's WPD field is absorbed by the blocking plate. The photon reaches the right-hand detector, generating an ×. Its second WPD field remnant travels to the upper detector, where it vanishes without a trace.

In Fig. 13-1(f), the remnant of the entering photon's WPD field is absorbed by the blocking plate. The photon reaches the upper detector, generating an ×. Its second WPD field remnant travels to the right-hand detector, where it vanishes without a trace.

This seemingly unremarkable Fig. 13-1(f) description is the raison d'etre for the Kwiat et al. article. If you erase (mentally, I trust) the WPD fields in Fig. 13-1(b) and Fig. 13-1(f), the following is what is left. In Fig. 13-1(b), a photon enters but is unrecorded. In Fig. 13-1(f), the blocking plate intercepts nothing at all, but the upper detector reveals the presence of a blocking plate by registering the arrival of a photon. The title of the article, "Quantum Seeing in the Dark," reflects the fact that the apparatus *somehow* "sees" the blocking plate even though no photon (that is, light) is actually intercepted by the blocking plate. The conjectured depiction of Fig. 13-1 says that the plate *does* block something, but it is a WPD field and not the photon that generated the field.

With the blocking plate, 25% of the entering photons follow the scenario shown in Fig. 13-1(f), registering an × in the upper detector. (The actual measurements have to be corrected for detector inefficiency.)

In Fig. 13-1(g), 50% of the entering photons are absorbed by the blocking plate and, therefore, do not reach a detector. Nevertheless, it is conjectured that WPD fields *do* reach the detectors, that they are compression shock waves in the aether, and that they are zero-energy fields that vanish without a trace.

The above analysis solves the "quantum seeing in the dark" mystery. The aether has been resuscitated!

# CHAPTER 14

▼

# RELATIVISTIC CHANGES

Another sensational experiment that strains quantum reality is one that involves electrons. Because of experimental difficulties, however, this was not successfully demonstrated until 1989 by A. Tonomura et al. [24]. One of the problems is that objects that have mass, such as electrons, become heavier and shorter as their velocity increases. Their velocity increases, that is, relative to the stationary, nonaccelerating observer who is making the measurements. Therefore, the changes in effective mass and length due to relative velocity are called *relativistic*. Their possible association with the aether is considered in Chapter 16.

The relativistic change in length is not pertinent to the discussion in the present chapter. Only the relative change in mass is considered.

There are three elementary particles that have mass: the electron, the proton, and the neutron. As given in Table A-2, their masses are, respectively, $9.1094 \times 10^{-31}$ kg, $1.67262 \times 10^{-27}$ kg, and $1.67493 \times 10^{-27}$ kg. Although the present chapter is concerned with objects that have mass, for convenience, only the electron is considered. Much of the discussion and conclusions, however, also apply to the proton and neutron.

"Massive" particles display gravitational attraction toward each other. However, this force is relatively weak. It is not pertinent to the present chapter.

If massive particles interact, momentum ($p$) is conserved. Momentum is given as

$$p = mv, \quad (14\text{-}1)$$

where $m$ is mass and $v$ is velocity. If we add up all of the $mv$ values of the particles *before* they interact, the sum has to equal the sum of $mv$ values *after* they interact. If it is a three-dimensional interaction, one must separately conserve momentum in the $x$, $y$, and $z$ directions.

It is interesting to contrast this with the "massless" photon. As pointed out in Chapter 1, photons ignore each other, and two photons that hit each other head-on only yield the algebraic sum of their respective wave packets. Following the "collision," they continue to propagate, unchanged, at the speed of light. This can be taken as further evidence for an aether carrier medium.

It is sometimes convenient, in this chapter, to consider the effective mass and momentum of a photon. The photon's effective mass is given as

$$m_{eff} = hf/c^2. \quad (14\text{-}2)$$

As one should expect, effective momentum is equal to effective mass times velocity. We get for a photon

$$p_{eff} = hf/c. \quad (14\text{-}3)$$

In connection with the photon generated when an electron spirals from the $n$ = 2 to the $n$ = 1 orbit of a hydrogen atom, a typical photon is emitted. When this happens, the following values are calculated: $m_{eff} = 1.819 \times 10^{-35}$ kilogram and $p_{eff} = 5.453 \times 10^{-27}$ kilogram·meter/second. These are extremely small values. The $m_{eff}$ is 50,000 times lighter than an electron, while $p_{eff}$ is 366 times smaller than that of the electron in the $n$ = 1 orbit. We can immediately conclude, there-fore, that even the lightest of "massive" particles, the electron, is a giant compared to a typical photon.

As luck would have it, we are surrounded by inexpensive equipment for examining electrons: cathode-ray tubes, also known as television picture tubes. A simplified model, without deflection plates or coils, is depicted in Fig. 14-1. In response to a positive voltage $V$, electrons are accelerated toward the fluorescent screen, striking it at high speed. Some of the electrons' kinetic energy is converted into light (photons). A permanent record of electron strikes can be obtained by placing a photographic film next to the fluorescent screen, as shown.

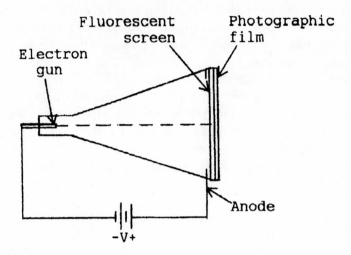

**Fig. 14-1.** Simplified model, without deflection plates or coils, of a cathode-ray tube. The photographic film provides a permanent record of electron strikes against the fluorescent screen.

The potential energy of the electric field is converted into kinetic energy as the electron speeds up. Using a "conventional" equation for the conversion, one finds that the speed of the electron is greater than the speed of light if $V$ is greater than 256,000 volts. This is, of course, impossible. Nothing, and certainly not a material object such as an electron, can travel faster than $3 \times 10^8$ m/s.

What is wrong? The "conventional" equation is at fault. As any material object increases in speed, its *effective mass* increases. The increase is such that, in converting potential energy to kinetic energy, the velocity of the object can never reach the speed of light.

The symbol $\gamma$ is used for the increase-in-mass ratio. For an electron,

$$m_{eff} = \gamma m_0, \quad (14\text{-}4)$$

where

$$\gamma = 1/[1 - (v/c)^2]^{1/2} \quad (14\text{-}5)$$

and $m_0$ is the electron's rest mass, $9.1094 \times 10^{-31}$ kg. Notice that, *not by coincidence*, $\gamma$ is equal to $t_2/t_1$ in Equation 8-3. The electron *behaves* as if it has a mass $\gamma m_0$ if it is moving, whether this is due to the $V$ of a cathode-ray tube or for any other reason.

## Some Numerical Values

Some of the numerical values that illustrate the above concepts are given in Table 14-1. The first column lists various voltages, $V$, applied to the cathode-ray tube anode with respect to its cathode. The second column lists the corresponding $\gamma$. The third column lists $v/c$ ratios for the velocity with which an electron strikes the fluorescent screen (using the correct, relativistic equation). The fourth column gives the correct, relativistic velocity. At relatively low voltages, the conventional and relativistic velocities are approximately the same; above $V = 25,000$ volts, however, one should only use the relativistic values.

**Table 14-1.** Various values associated with an electron as potential energy $eV$ is converted into kinetic energy $K$. $\gamma$ is the relativistic increase in mass factor; $v/c$ and $v$ include the relativistic effect; $f_{PWD}$ and $\lambda_{PWD}$ are the frequency and wavelength of the particle-wave duality field. Because this is *not* an electromagnetic field, the last column is for identification only; no orange, ultraviolet, X-ray, or gamma-ray energy is actually available.

| $V$ (volts) | $\gamma$ | $v/c$ | $v \times 10^8$ (m/s) | $f_{PWD}$ (Hz) | $\lambda_{PWD}$ (Å) | ID |
|---|---|---|---|---|---|---|
| 1 | 1.000 | 0.00198 | 0.00593 | $4.836 \times 10^{14}$ | 12.26 | Orange |
| 10 | 1.000 | 0.00626 | 0.01876 | $4.836 \times 10^{15}$ | 3.878 | Ultraviolet |
| 100 | 1.000 | 0.01978 | 0.05930 | $4.836 \times 10^{16}$ | 1.226 | Ultraviolet |
| 1000 | 1.002 | 0.06247 | 0.1873 | $4.831 \times 10^{17}$ | 0.3876 | X-ray |
| 10,000 | 1.020 | 0.1950 | 0.5846 | $4.790 \times 10^{18}$ | 0.1220 | X-ray |
| 25,000 | 1.049 | 0.3018 | 0.9049 | $1.181 \times 10^{19}$ | 0.07664 | X-ray |
| 50,000 | 1.098 | 0.4127 | 1.237 | $2.310 \times 10^{19}$ | 0.05355 | X-ray |
| 100,000 | 1.196 | 0.5482 | 1.644 | $4.440 \times 10^{19}$ | 0.03701 | X-ray |
| 510,990 | 2 | 0.8660 | 2.596 | $1.853 \times 10^{20}$ | 0.01401 | $\gamma$-ray |
| $10^6$ | 2.957 | 0.9411 | 2.821 | $3.236 \times 10^{20}$ | 0.008719 | $\gamma$-ray |
| $10^7$ | 20.57 | 0.9988 | 2.994 | $2.536 \times 10^{21}$ | 0.001181 | $\gamma$-ray |
| $10^8$ | 196.7 | 1.0000 | 2.998 | $2.430 \times 10^{22}$ | 0.000123 | $\gamma$-ray |

Incidentally, $V = 25,000$ volts is typical for a cathode-ray tube (but the current is very small). Notice that the electron reaches a very impressive velocity, $0.9049 \times 10^8$ m/s (56,000 miles/s). No wonder the fluorescent screen lights up!

An important level of $V$ occurs at 510,990 volts, which corresponds to $\gamma = 2$. This value of $V$ is the basis for a convenient unit of electron mass because 510,990 electron volts/$c^2$ is equal to $m_0$.

A photon behaves as if it has an effective mass and effective momentum. In this way, the photon, which is an electromagnetic wave packet, displays the characteristics of a particle that has mass. In 1924, Louis de Broglie proposed that the reverse may be true: that an electron, which has mass, can display the characteristics of a wave. Soon afterward, experiments showed that de Broglie's hypothesis was correct; in fact, every mass in motion, in general, demonstrates wave characteristics. De Broglie's conjecture was an important milestone that was recognized by a Nobel Prize in 1929; besides, because it was made via a relatively short PhD thesis, it has fired the imagination, if not inspiration, of every PhD physics student since 1924.

The reason for considering the massive electron versus the wavelike photon is that each of them displays an interference pattern in the double-slit apparatus. They are, however, two different species. The electron's field travels at the speed of the electron, which can be anything from zero up to the upper limit, the speed of light, while the photon's WPD field *always* travels at the speed of light (in a vacuum). Also, the frequency of the electron's field is a function of its velocity, while the photon's WPD field frequency is that of its power pack. Therefore, in what follows, the electron's field is called a particle-wave duality (PWD) field to distinguish it from the photon's WPD field.

The particle-wave duality (PWD) frequency of an electron is given by the photon's $m_{\text{eff}} = hf/c^2$ if we substitute the electron's velocity $v$ in place of the photon's velocity $c$. This yields

$$f = m_{\text{eff}} v^2 / h, \quad (14\text{-}6)$$

where $h = 6.6261 \times 10^{-34}$ joule·second, Planck's constant. However, since the electron's effective mass is a function of velocity, it is more convenient to substitute $m_{\text{eff}} = \gamma m_0$ to get

$$f = \gamma m_0 v^2 / h. \quad (14\text{-}7)$$

This is the equation used to calculate values in the frequency column, $f_{\text{PWD}}$, of Table 14-1. Wavelength, $\lambda_{\text{PWD}}$, is given by velocity/frequency, as usual. Because the PWD field is *not* an electromagnetic field, the last column is for identification only; no orange, ultraviolet, X-ray, or gamma-ray energy is actually available.

# CHAPTER 15

▼

# THE DOUBLE-SLIT ELECTRON
# INTERFERENCE PATTERN

The particle-wave duality (PWD) frequency values in Table 14-1 are relatively high. As mentioned previously, an electron is a giant compared to a photon, and this shows up in the associated frequency values. At a typical cathode-ray tube value of $V$ = 25,000 volts, Table 14-1 shows $f$ = 1.181 × 10$^{19}$ Hz. According to the table, this is an X-ray frequency, as is indicated in the last column of Table 14-1. I hasten to add that these are *not* the X-rays that, it is frequently claimed, are emitted by a cathode-ray tube. The electron's particle-wave dual is an X-ray in frequency only; it is not an electromagnetic wave; it propagates at the velocity of the electron, not that of light; it has zero energy, zero penetrating power, and vanishes without a trace when the electron strikes its fluorescent screen. Is it realistic for us to believe that this field has zero energy? The arguments regarding the energy of the photon's zero-energy WPD field, in Chapter 12, apply equally well to the electron's PWD field.

The bona fide X-rays that the screen *does* emit are due to the great velocity with which an electron arrives at the screen. Part of the kinetic energy is converted into fluorescent excitation; part of the kinetic energy is converted into photons in the X-ray range of frequencies; and part of the kinetic energy is converted into heat. In the case of a television receiver, it is generally considered that the X-ray effect is negligibly small, especially compared to that of the deadly program material.

Nevertheless, the high PWD frequencies offer almost insurmountable experimental difficulties when attempting to demonstrate the incontestable signature of a wave: constructive and destructive interference in the double-slit apparatus. It is interesting to consider, below, how some of the difficulties were overcome.

The proof that an electron can act as a wave came from the same techniques that are used to prove that an X-ray is a wave. For example, the above-mentioned $V$ = 25,000-volt PWD frequency has a wavelength of 0.077 angstrom (Å). In Fig. 11-1, the spacing between the two slits is around 5 wavelengths, so a spacing of 0.4 Å would be reasonable for the electron beam. The "slits" in this case can be provided, many of them, by the repetitive spacing between the atoms of a crystalline material. Clinton Davisson and Lester Germer, in 1925, showed electron diffraction and interference using a crystal made out of nickel.

In 1989 the "impossible" was accomplished: five physicists, A. Tonomura et al., used skill, persistence, ingenuity, and modern equipment to demonstrate the particle-wave duality of electrons [16]. In what follows, I am going to take advantage of the accomplishment of Tonomura et al. by using the double-slit photon interference drawings of Chapter 11 and applying them to double-slit electron interference in this chapter. Changes in the text and drawings of Chapter 11 are made, as needed, to accommodate electrons rather than photons.

As an electron source, Tonomura et al. used a sharp field-emission tip and an anode potential of 50,000 volts. According to Table 14-1, $f$ and $\lambda$ were $2.3 \times 10^{19}$ Hz and 0.054 Å. Tonomura et al. state that "[w]hen a 50-kV electron hits the fluorescent film, approximately 500 photons are produced from the spot." They used a much more sophisticated light-gathering arrangement, including a magnification of 2000, than the photographic film shown in Fig. 14-1.

For electrons, one must employ a high vacuum, in addition to facing the problems associated with angstrom-size wavelengths. As a vehicle for this discussion, consider the idealized double-slit interference-diffraction apparatus of Fig. 15-1(a). The electron beam is moving to the right. It strikes a plate that contains two slits. Some of the electrons get through the upper slit, and some get through the lower slit. To the right of the slits, the electrons spread out, via diffraction, as if they had wave characteristics. Two of the rays thus formed, (1) of length $e_1$ and (2) of length $e_2$, are shown as they come together on a fluorescent screen. A relatively high voltage is used so that the electrons will have sufficient energy to elicit a fluorescent response that can be recorded on the film. What pattern will the exposed film show?

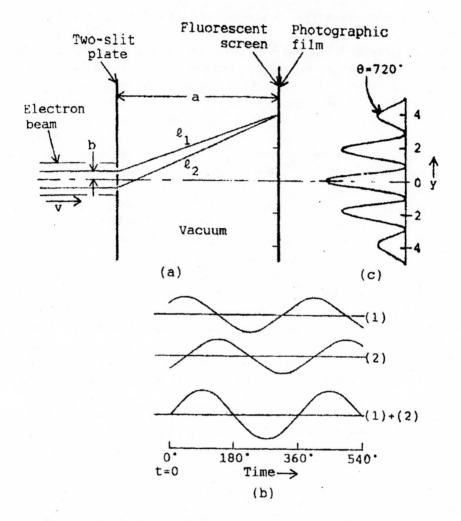

**Fig. 15-1.** Double-slit interference and diffraction. (a) This shows a schematic of an idealized apparatus based on the fact that Tonomura et al. have demonstrated the particle-wave duality (PWD) of electrons [24]. The slits are at right angles to the page. Two of the rays leaving the slits are depicted as they meet at $y = 4$ on the fluorescent screen. (b) This shows waveforms of rays (1) and (2) when they meet at the screen if they are 90° out of phase. (c) This shows the film pattern. The *same* pattern results if the input "beam" consists of single, one-at-a-time electrons.

In some locations, the PWD field from ray (1) is in phase with that of ray (2) when they meet at the screen, and the electrons associated with the PWD fields increase film exposure (constructive interference). At other locations, the PWD fields have opposite phases, and the electrons avoid these regions (destructive interference). Fig. 15-1(b) illustrates an in-between situation in which the PWD fields are 90° out of phase. The net results of constructive and destructive interference are the idealized set of peaks and valleys shown in Fig. 15-1(c).

Fig. 15-1 depicts the following relative values: $b = 0.5$, $a = 10$, $y = 4$, and $\theta = 720° = 4\pi$ (the second peak away from the $y = 0$ axis). The numerical values correspond to a PWD field relative wavelength of $\lambda = 0.1857$. Ray $\ell_2$ is 10.97 units long and contains 59 cycles of PWD field. Ray $\ell_1$ is 10.59 units long and contains 57 cycles. Therefore, the two signals arrive in phase (constructive interference).

At $y = 1$ ($\theta = 180° = \pi$), the longer path is 54.4 cycles long, and the shorter path is 53.9 cycles long, so the difference is 0.5 cycle. Therefore, the two signals cancel (destructive interference).

The film pattern follows a commonly encountered $[(\sin y)/(y)]^2$ intensity pattern.

Now consider that the electron beam carries kinetic energy. If an electron does not arrive at the screen because of destructive interference, it must be picked up by regions of constructive interference. After an electron gets past the double slits, it diffracts by an amount that is based on its predetermined, but statistically random, prior experiences. Then, because of the bending of pseudo-$\underline{E}$ lines, entering electrons veer off toward the upper and lower paths, avoiding the middle destructive-interference path. In other words, in Fig. 15-1(a), the electrons actually curve away from destructive-interference points $y = \pm1$ and $y = \pm3$, and toward constructive-interference points $y = 0$, $y = \pm2$, and $y = \pm4$. As a result of this "curving away," the valleys of Fig. 15-1(c) are created.

### Simultaneous-Burst Pattern

Our next step is to carefully decrease the output of the electron beam. Suppose that an ideally fast pulse allows a burst of only 1000 electrons to *simultaneously* fly through the slits. The procedure used to derive Fig. 11-2(b), for a photon beam, was applied to the film exposure display of Fig. 15-1, yielding Fig. 15-2(b). Here, out of the 1000 electrons, 150 end up in the $y = 0$ bin. This is reasonable if half of the $y = 0.5$ and $y = -0.5$ electrons, from Fig. 15-2(a), are captured by the $y = 0$ bin. But what happens to the 71 electrons that, according to Fig. 15-2(a), started out headed for $y = 1$? Fig. 15-2(b) tells us that only 2 get through. What happens to the other 69 electrons? They end up in the constructive-interference regions to either side of $y = 1$.

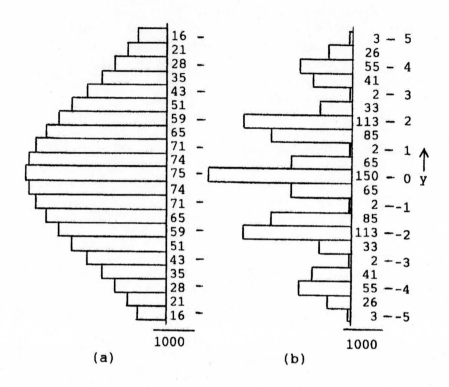

**Fig. 15-2.** Electron exposure distributions at the screen-film of Fig. 15-1(a) if the bins are 0.5$y$ unit wide. (a) This shows the electron distribution, due to an assumed diffraction attenuation function, $\exp(-0.0625y^2)$, with the interference effects omitted. (b) This shows the electron distribution including constructive and destructive interference, as in Fig. 15-1(c).

## Individual-Electron Pattern

Finally, instead of 1000 simultaneous electrons, we restrict the beam so effectively that *only one isolated electron at a time gets through*—one per second, say. After 1000 seconds, we develop the film. We expect to see Fig. 15-2(a) because constructive or destructive interference could not possibly occur with individual electrons. Instead, we get Fig. 15-2(b)! This is an unbelievable result, impossible to explain by classical physics or realistic quantum physics.

The Tonomura et al. paper is titled "Demonstration of Single-Electron Buildup of an Interference Pattern." In its entirety, their abstract follows [24]:

> The wave-particle duality of electrons was demonstrated in a kind of two-slit interference experiment using an electron microscope equipped with an electron biprism and a position-sensitive electron-counting system. Such an experiment has been regarded as a pure thought experiment that can never be realized. This article reports an experiment that successfully recorded the actual buildup process of the interference pattern with a series of incoming single electrons in the form of a movie.

The Tonomura et al. experiments show that the film exposure display of Fig. 15-1(c) occurs independent of electron beam density. Their paper reproduces five film exposures, showing how the electron interference pattern builds up as the number of *individual* electrons striking the fluorescent screen increases: 10, 100, 3000, 20,000, and 70,000. In my opinion, this illustration is one of the most remarkable in the history of science. The 70,000-electron film exposure looks exactly like Fig. 11-3 except for a difference in contrast and spacing.

The evidence would have us believe that an electron somehow divides in half, and each half goes through a slit. Upon emerging from the slit, each half is apparently associated with an EMF that is similar to that of 1000 simultaneous electrons (except that the total EMF energy is that of a single electron). The emerging EMFs cover the entire screen of Fig. 15-1(a), from $y = -5$ to $y = +5$. The energy of the EMF that strikes the screen should be modified by constructive and destructive interference, as depicted in Fig. 15-1(c). Instead, the electron behaves like a point particle, lands on the screen at $y = 4$, say, and *all* of its energy is converted into a single bright dot at $y = 4$. After 1000 seconds, it will turn out that some 55 electrons [a value given by Fig. 15-2(b)] were captured by the $y = 4$ bin; 150 landed in the $y = 0$ bin; and so forth.

There are two serious problems with the above recital. First, since an electron is an irreducible constituent of matter, it cannot split into two halves, each passing through one of the slits. Second, if the electron gives birth to an EMF-type field that covers the entire screen from $y = -5$ to $y = +5$, the electron's energy would reside in this field, leaving less than a normal amount for the particle that eventually strikes and stimulates fluorescence at $y = 4$.

# CHAPTER 16

▼

# THE PARTICLE-WAVE DUALITY FIELD

In what follows, I propose that the electron is accompanied by a PWD field that is similar to the photon's WPD field. "Similar," but different in two major respects: (1) the electron and its entourage can move at any velocity *less than c*, whereas a photon propagates *at* velocity $c$ (through a vacuum); and (2) the frequency of the electron's PWD field is a function of $v$, given by Equation 14-7, $f = \gamma m_0 v^2 / h$, whereas the frequency of the photon's field is equal to that of its power pack (the wave packet). Because of these differences, it is necessary to classify the two fields as belonging to altogether different species.

My conjecture here is that the electron's PWD field is a type of compression *wind* generated as the electron flies through the aether. (Although it is nominally a "compression" wind, it actually consists of compressions and expansions). This is analogous to air versus a low-speed projectile, such as a pitched baseball. In the aether, the PWD field corresponds to compressions and expansions that precede the power pack. It is again conjectured that these ethereal waves do not convey any energy.

My argument regarding zero energy paraphrases the discussion in Chapter 12. There is no attenuation of electrons in a vacuum, provided we restrict it to a special case: the vacuum must not contain $\underline{E}$ or $\underline{H}$ fields, since the electron may interact (accelerate or decelerate) if these fields are present. In the absence of $\underline{E}$ or

-115-

H fields, an electron travels in a straight line, at constant speed, in a vacuum (if gravity is ignored). A change in speed implies the conversion of electron energy into heat, which in turn implies that some particle, such as an atom, will vibrate more rapidly as it absorbs this energy. But there are of course no atoms in our vacuum; it consists of aether and nothing else, so the PWD fields have to be zero-energy fields. The aether, if it exists, is a perfectly elastic, lossless, linear medium.

Inside the electron's "power pack" is a negative charge, $e = 1.6022 \times 10^{-19}$ coulomb, mass $m_0 = 9.1094 \times 10^{-31}$ kilogram, and normalized spin $s = 1/2$. The spin of a particle is its angular momentum that exists even when the particle is at rest, just as it has a mass $m_0$ at rest. (Here we *can* think of the particle as if it were a minuscule spinning baseball. The spin of an electron enters into the discussion in Chapter 17.) Visualize an electron as flying off to the right at velocity $v$; preceding it is the PWD field whose frequency and wavelength are given by $f = \gamma m_0 v^2 / h$ and $\lambda = v/f$.

The picture that emerges is this: an electron at rest has a negative charge $e$, mass $m_0$, and spin $s$. As soon as it starts to move, a PWD field develops. For example, when it has converted 1 volt into kinetic energy, Table 14-1 tells us that the electron model is moving (to the right, say) at a velocity of 593,000 meters/second (1,326,000 miles/hour). This is relatively slow for an electron! The PWD field lines are 6 angstroms apart between the positive and negative peaks. (The wavelength is 12 Å.) The lines zoom by at a frequency of $4.836 \times 10^{14}$ Hz. Although this corresponds to an orange glow, there is of course no visible effect when the electron strikes the double-slit plate. The PWD field, to repeat, is not an EMF, and probably carries zero energy.

As an electron accelerates, frequency increases and the wavelength shrinks. At 25,000 volts, $f = 1.18 \times 10^{19}$ hertz and $\lambda = 0.077$ angstrom. Beyond this voltage, relativistic effects become appreciable; the electron behaves as if its mass is increasing in accordance with $\gamma m_0$. At a potential of 510,990 volts, $\gamma = 2$, the PWD frequency is $1.85 \times 10^{20}$ Hz, and $\lambda = 0.014$ Å.

The experiment of Tonomura et al. shows that the PWD really exists. The electron interference pattern is there, literally in black and white. Their pattern agrees with the 50,000-volt calculated wavelength of 0.054 angstrom. At this voltage, relativistic effects are also verified since $\gamma$ is appreciably greater than 1 (it is 1.10).

Is the PWD field longitudinal, like a sound wave, or transverse? For a photon, polarization shows that the WPD field is transverse. If the PWD field is a compression wind wave in the aether, however, it is analogous to a longitudinal wind disturbance in air, and the polarization plane becomes meaningless.

The changes in effective mass and PWD wavelength occur because the electron is moving. With respect to what? With respect to the electron gun in a cathode-ray tube in a physics laboratory? What about the relativistic effect? According to $E = mc^2$, the effective mass is proportional to the energy carried by the electron. Are we prepared to say that an observer moving with the electron (as it drifts at constant speed past the anode, say) will see no change in mass and no PWD field? As the electron flies through the aether, a "viscosity" interaction induces wind waves (the PWD field). It seems to me that it is much easier to visualize this change if an aether is present.

So a photon can travel through the aether at the speed of light, without attenuation, whereas an electron runs up against an aether that has effective mass. These are, indeed, strange conjectures.

How far does the PWD field extend in front of the power pack? At least 10 or 20 wavelengths, enough to get a reasonably effective degree of destructive interference.

Views (a) through (g) of Fig. 16-1 now depict how the electron model can explain the double-slit experimental results for a single, isolated electron (Fig. 15-1). The text would follow almost word-for-word the photon discussion of Chapter 14 in connection with Fig. 12-1. The caption of Fig. 16-1 is sufficiently detailed to serve as the text with a minimum of further explanation.

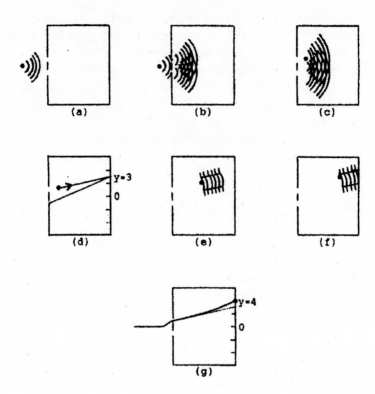

**Fig. 16-1.** Sequence that illustrates double-slit interference effects that accompany a single, isolated electron. (a) This shows an electron approaching the slit plate. (b) This shows that the leading portion of the PWD field has split, with a fragment getting through each of the slits. (c) This shows that the PWD fields have progressed beyond the slit plate. The power pack, because of its predetermined but statistically random past history, has followed the upper-slit PWD segment. (d) This shows the same as (c), but with PWD fields omitted. The power pack is heading for the $y = 3$ point of the screen. (e) This shows the power pack and the net PWD field halfway across. (f) This shows that, because the PWD field lines are concave, the power pack is directed away from the destructive-interference $y = 3$ point. (g) This shows the power pack locus curves striking the screen at the $y = 4$ point. The ethereal PWD field has vanished without a trace.

With regard to the lateral force needed to change the electron's direction in going from Fig. 16-1 (b) through (c), it is again conjectured that the aether forms streamlines through the two slits. These guide or steer the electron as the aether supplies the lateral force that is required. There is no change in kinetic energy if no change in speed is involved, so the lateral push need not entail a change in energy.

In Fig. 16-1(e), the power pack is midway between the double-slit plate and the fluorescent screen. Because it is approaching a destructive-interference point, the PWD field lines are concave. This "encourages" the power pack to head for the constructive-interference points at $y = 2$ or 4.

Fig. 16-1(g) shows the path taken by the power pack. The PWD field is an ethereal compression wind; it vanishes without a trace.

Tonomura et al. do not attempt to explain the unrealistic experimental outcome. The statistical predictions of quantum mechanics are of no help here because we are dealing with the interference pattern associated with a *single* electron.

## Comment

What do physicists have to say about the single-electron results? Again, with the kind permission of Springer Science and Business Media, I would like to quote a physicist. In physicist Tore Wessel-Berg's book, *Electromagnetic and Quantum Measurements: A Bitemporal Neoclassical Theory*, he states the following [26] (pp. 205–206):

> Together with the double slit experiment for photons the corresponding experiment for electrons represents one of the most celebrated and famous experiments in physics. Its fame is due more to the conceptual problems it creates rather than its contribution to understanding quantum physics. The observation of wavelike diffraction of electrons and their definite particlelike behavior in other circumstances has remained a conceptual mystery in quantum physics until this day. The basic experimental setup...is simple enough. A stream of focused electrons from a hot cathode impinges on a plate with two narrow slits separated a small distance apart. The electrons transmitted through the slits are observed to form a typical diffraction pattern on the screen behind the slit plate. If the intensity of the electron source is reduced to the point when only one electron at a time is reaching the screen, it produces a pointlike spot located somewhere on the screen, not necessarily just below the slits. This behavior is certainly in accord with classical concepts of the electron

as a particle, with the electron passing through *one* of the two slits. In the process it is deflected by some angle and finally hitting the screen at some localized point. The deflection has no classical explanation. But there are worse things to come. As more and more electrons are arriving at the screen the overall average macroscopic intensity builds up to a typical diffraction pattern as indicated in the figure. The similarity of the pattern with the familiar diffraction of an optical beam impinging on a double slit plate is immediately apparent.

In the absence of an immediate classical explanation quantum physics took refuge in the concept of duality. According to this principle the electron can be both a particle and a wave, depending on circumstances. When the single electron approaches the slit plate it mysteriously transforms itself to a wave encompassing both slits. The two wave components originating from the slits interfere behind the slit plate to produce the diffraction pattern, in accordance with regular wave principles. On hitting the screen the electron again transforms itself from a wave to a particle. The only virtue of such an explanation is that it explains the diffraction pattern. But it creates a number of questions regularly referred to as puzzles or paradoxes. One of the most famous is the "which way" paradox. Classically, the electron must pass through *one* of the slits. But the diffraction pattern can not be explained unless the electron passes through *both* slits. If one tries to somehow measure which slit the electron passes through, the diffraction pattern disappears. It appears that nature refuses any attempt to gain information on its secrets, responding with severe penance in the form of experiment failure.

Again, I point out that Professor Wessel-Berg is *not* a "mainstream physicist" who does not have an explanation for the quantum paradoxes. Please see his "bitemporal neoclassical theory."

# CHAPTER 17

▼

# AN ELECTRON-SPIN EXPERIMENT

This chapter continues with the consideration of an experimental setup that yields a result that cannot be explained by any existing reality, but which can be explained by the PWD concept. I am referring to Fig. 17-1, which is discussed by David Z. Albert [33].

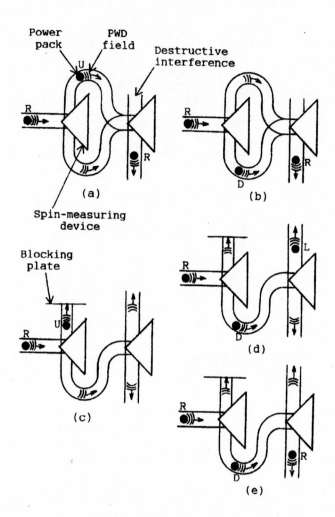

**Fig. 17-1.** Experiment that yields strange results [33], but that can be explained by the PWD model. R, L, U, and D are right-spinning, left-spinning, up-spinning, and down-spinning electrons, respectively. Each spin-measuring device causes a 90° change in spin direction. All of the input electrons have preselected R spins, but 50% develop U spins, and the other 50% develop D spins. (a) This shows what happens if an entering electron develops a U spin. (b) This shows what happens if an entering electron develops a D spin. In (c), (d), and (e), the U output of the first spin-measuring device is blocked by a plate. (c) This shows the same as (a) except there is a blocking plate. (d) This shows the same as (b) except there is a blocking plate and the D electron develops an L spin. (e) This shows the same as (b) except there is a blocking plate and the D electron develops an R spin.

Here we have two identical devices that measure electron spin; they are represented by triangles. The spin-measuring function is not important here; instead, observe that the triangles cause a 90° change in spin direction. If a right-spinning (R) electron enters the first triangle, it comes out either as an up-spinning (U) or down-spinning (D) electron. If a U or D electron enters the second triangle, it comes out either as an R electron or as a left-spinning (L) electron.

All of the electrons fed into the equipment have preselected R spins, but 50% develop U spins, and the other 50% develop D spins. They enter individually, one at a time.

Fig. 17-1(a) illustrates the scenario in which 50% of the electrons develop U spins. When an R electron enters the first triangle, its PWD field splits, half taking the up path and, simultaneously of course, half taking the down path. The power pack, based upon its past history, takes the U path. Both paths are brought together with the aid of reflectors (not shown), which, say, cause the electron loci to again become horizontal. When the U electron enters the second triangle, a strange effect results: only R electrons leave the triangle. Apparently, the phase relationships are such that destructive interference between the upper and lower PWD field branches occurs, so that there is no L output. Instead, with the aid of constructive interference, all of the power packs emerge as R electrons.

Fig. 17-1(b) illustrates the scenario of the 50% that develop D spins. As before, when an R electron enters the first triangle, its PWD fields split. This time the power pack, based upon its past history, takes the D path. When this D electron enters the second triangle, only R electrons again leave the triangle. Apparently, the phase relationships are such that destructive interference between the upper and lower PWD field branches again occurs, so that there is no L output.

Fig. 17-1(c), Fig. 17-1(d), and Fig. 17-1(e) depict the outcomes if a blocking plate is placed over the U output of the first triangle. Fig. 17-1(c), like Fig. 17-1(a), illustrates the scenario of the 50% that develop U spins. These power packs strike the blocking plate, where their kinetic energy is converted into heat. The lower PWD field splits, and the two segments leave the second triangle as shown. These are zero-energy fields that vanish.

Fig. 17-1(d), like Fig. 17-1(b), illustrates the scenario of the 50% that develop D spins. Now, because of the blocking plate, another strange effect results: with only one PWD field, interference cannot take place. Now there *is* an L output, as shown. This is the path taken by half of the D electrons that enter the second triangle; the other half take the R output path, as shown in Fig. 17-1(e).

To summarize, *without* the blocking plate, 100% of the entering electrons leave as R electrons; *with* the blocking plate, 50% are absorbed by the plate, 25% leave as R electrons, and 25% leave as L electrons. It is explained by constructive and destructive interference as the zero-energy PWD fields interact. Remember, however, the basic conjectures: PWD fields are compression (nominally) wind waves in the aether, and electrons tend to be guided by streamlines in the aether, which can also supply lateral forces.

Again, the particle-wave duality field model solves the mystery and provides support for the resuscitation of the aether.

# CHAPTER 18

▼

# A FASTER-THAN-LIGHT EXPERIMENT

In certain experiments involving pairs of photons, to be described below, it appears as if an action visited upon one of the photons is instantaneously felt by the other photon, even if it is relatively far away. This is known as "entanglement." John S. Bell pointed out that the correlation between the two photons exceeded the expectation allowed by a local (speed of light) phenomenon [34, 35]. Bell's theorem states that certain experimental results *must* be non-local; that is, they display superluminal (faster-than-light) behavior. In this chapter, one of the representative experiments is considered.

But superluminal transmission of information is strictly forbidden in electromagnetic field theory, as well as by common sense. According to Bell, if the experimenter imparts a change to photon C, it can almost instantaneously cause a corresponding change to photon D, millions of meters away. It appears as if an explanation requires conjectures that bypass quantum mechanics.

The experiment is based on the block diagram of Fig. 18-1. Here the central block is a twin-state photon generator. There are several ways to generate a *single pair* of photons. For many elements, if the atoms are placed into an excited state, their outer electrons emit a pair of photons when they return to their ground state (in contrast with a hydrogen atom, where the single electron can only launch a single photon). Favored sources include mercury excited by an electron beam, and calcium excited by a laser beam. The two emitted photons have differ-ent frequencies: for the calcium cascade, we have $f = 5.438 \times 10^{14}$ Hz (yellow-green) and $f = 7.092 \times 10^{14}$ Hz (indigo-violet). The different colors are of little consequence. The important aspect of twin-state emission is that both photons have the same polarization angle, $\phi$, as indicated in Fig. 18-1. (Actually, because the net angular momentum of the photon pair must be zero, the photons are emitted with opposite angular momentums, but this translates into the same value of $\phi$ insofar as the experiment is concerned.)

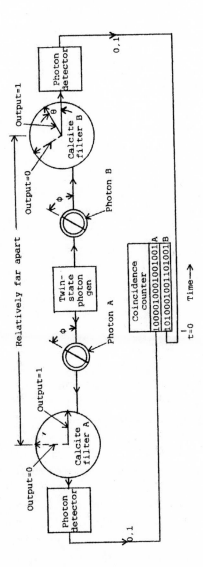

**Fig. 18-1.** Experiment from which it is concluded that photons A and B somehow communicate with each other superluminally. The photons are simultaneously emitted by the twin-state photon generator; although their polarization angle $\phi$ is a random variable, $\phi$ is the same for both photons. The calcite filter angles are set to $\theta_A = 0$ on the left and, manually, to $\theta_B = \theta$ on the right. Depending on $\phi$ versus $\theta_A$ or $\theta_B$, respectively, each photon exits the filter either along the 1 or 0 output paths. A coincidence counter keeps track of simultaneous matches between A and B outputs.

The actual polarization angle *varies randomly* from $-90°$ to $+90°$, but left and right photons have the *same* $\phi$.

Eventually, after traveling a relatively large distance, the photons enter calcite filters A and B. The distance is "relatively large" in the sense that the time taken for a signal to travel between A and B is appreciable, even at the speed of light. For example, a distance of 3 meters requires 10 nanoseconds, but this time is appreciable, and it can be measured easily with a sophisticated electronic clock.

## Calcite Filters

Shine a small-diameter ray of visible light onto the surface of a slab of glass. (See Fig. 7-5.) Let the angle between the ray and normal (perpendicular line) to the surface be $\theta_1$. Going from air into glass, the speed of light is reduced; this causes a ray to bend so that, in the glass, $\theta_2$ is less than $\theta_1$. (The ray bends toward the normal.) This action is described by Snell's law:

$(\sin \theta_1)/(\sin \theta_2)$ = (velocity in medium 1)/(velocity in medium 2).

If medium 1 is air (or a vacuum), the velocity in medium 1 is $c \cong 3 \times 10^8$ m/s, and the velocity ratio is called the index of refraction. The index is always greater than 1.

If, instead of glass, we use an anisotropic, birefringent material, such as calcite (calcium carbonate), something unusual occurs. As "birefringent" implies, the calcite has *two* indices of refraction. In general, two rays, corresponding to two different values of $\theta_2$, form at the interface between, say, air and calcite. There is more to it than this, however. It turns out that, if the polarization of one of the internal rays is horizontal, say, then the polarization of the second ray is vertical. The polarization angle between the two rays always has a difference of $90°$.

Recall that a ray of light is an electromagnetic field (EMF) with $\underline{E}$ and $\underline{H}$ lines perpendicular to each other and to the direction of propagation, as illustrated in Fig. 4-1. The polarization angle is determined, by definition, by the direction of the $\underline{E}$ lines. Therefore, shine a small-diameter ray onto the surface of a slab of calcite. Let the angle between the ray and normal to the surface be $\theta_1$. Inside the calcite, two rays form. We can orient the slab so that $\theta_{2H}$ is a horizontally polarized component, while $\theta_{2V}$ is a vertically polarized component.

The applied (incident) ray, in general, splits into two vector components, inside the calcite, to form $\theta_{2H}$ and $\theta_{2V}$ rays. If the applied ray is horizontally polarized, all of it will (ideally) form the $\theta_{2H}$ ray, leaving nothing for the $\theta_{2V}$ ray, and so forth.

Next, we carry on in the tradition set forth in Chapter 11. We block the light source so effectively that only one isolated photon at a time gets through. Since the photon is the irreducible constituent of an EMF, it cannot split into two vector components. How does the photon avoid a split personality? It will probably follow the path that is *closest* to its angle of polarization.

To illustrate with numerical values, using standard four-quadrant angle orientation, let the polarization of the input photon be $\phi$, where $\phi$ lies between $-90°$ and $+90°$. (Angles outside of this range can always be extended into this range. For example, $120°$ is the same as $-60°$, $-135°$ is the same as $+45°$, and so forth.) Then some simple sketches will show that

if $\phi$ lies between $-45°$ and $+45°$, the internal ray follows the $\theta_{2H}$ path;

if $\phi$ lies between $+45°$ and $+90°$, the internal ray follows the $\theta_{2V}$ path; and

if $\phi$ lies between $-90°$ and $-45°$, the internal ray follows the $\theta_{2V}$ path.

In Fig. 18-1, to avoid confusion regarding H and V rays when they are not actually horizontal and vertical, the H and V subscripts are abandoned. Instead, one internal ray is shown as a solid line and its output is labeled "1"; the other is a dashed line and its output is labeled "0." The solid-line direction for calcite filter A is along the x axis ($\theta_A = 0$), but B's direction is a manually adjustable angle, $\theta_B$. Therefore, in what follows, the calcite *difference* angle, $\theta = \theta_B - \theta_A$, is equal to $\theta_B$.

The manually adjustable $\theta$ can range from $0°$ to $90°$, while the incoming polarization angle, $\phi$, can range from $-90°$ to $+90°$, as noted previously. The path that the internal ray follows (1 or 0) depends upon the difference, $\phi - \theta$. This can range from

$$\phi = -90°, \theta = 90°, \text{ so that } \phi - \theta = -180° \text{ to}$$
$$\phi = +90°, \theta = 0°, \text{ so that } \phi - \theta = 90°.$$

This can become very confusing because of the mixture of positive and negative values, but it turns out that the *magnitude* of $\phi - \theta$, or $|\phi - \theta|$, is the important variable because the internal ray "will probably follow the path that is *closest* to its angle of polarization." Here is the revised set of rules (notice the absence of negative values) given by $|\phi - \theta|$:

if $|\phi - \theta|$ lies between $0°$ and $45°$, the internal ray follows the 1 path;

if $|\phi - \theta|$ lies between $45°$ and $135°$, the internal ray follows the 0 path; and

if $|\phi - \theta|$ lies between $135°$ and $180°$, the internal ray follows the 1 path.

In Fig. 18-1, $\phi = 60°$ and $\theta_A = 0°$, so $|\phi - \theta| = 60°$ and filter A's output is 0. For filter B, $\phi = 60°$ and $\theta_B = \theta = 30°$, so $|\phi - \theta| = 30°$ and filter B's output is 1.

In Table 18-1, $\phi$ goes from $-82.5°$ to $+82.5°$ as $\theta$ goes from $0°$ to $90°$. The only way for you to get unconfused is to check some of my answers (which are, of course, never wrong). The table gives the outputs (0 or 1) of calcite filters A and B, and also a matching value, M = 1, if the filter outputs are the same.

**Table 18-1.** Expected coincidence counter matches in Fig. 18-1 as the photons' polarization angle, φ, takes on values between −82.5° and +82.5° while the calcite filter difference angle is set for values between 0° and 90°. The A columns represent the output of the A filter, which remains the same for the entire table because it is not rotated. The B columns, however, rotate down one row distance, φ = 15°, as we move to the right one θ column distance, θ = 15°. The M columns list the number of matches.

| Photon polarization angle φ | Filter difference angle, θ | | | | | | |
|---|---|---|---|---|---|---|---|
| | 0° ABM | 15° ABM | 30° ABM | 45° ABM | 60° ABM | 75° ABM | 90° ABM |
| −82.5° | 0 0 1 | 0 0 1 | 0 0 1 | 0 0 1 | 0 1 | 0 1 | 0 1 |
| −67.5° | 0 0 1 | 0 0 1 | 0 0 1 | 0 0 1 | 0 0 1 | 0 1 | 0 1 |
| −52.5° | 0 0 1 | 0 0 1 | 0 0 1 | 0 0 1 | 0 0 1 | 0 0 1 | 0 1 |
| −37.5° | 1 1 1 | 1 0 | 1 0 | 1 0 | 1 0 | 1 0 | 1 0 |
| −22.5° | 1 1 1 | 1 1 1 | 1 0 | 1 0 | 1 0 | 1 0 | 1 0 |
| −7.5° | 1 1 1 | 1 1 1 | 1 1 1 | 1 0 | 1 0 | 1 0 | 1 0 |
| 7.5° | 1 1 1 | 1 1 1 | 1 1 1 | 1 1 1 | 1 0 | 1 0 | 1 0 |
| 22.5° | 1 1 1 | 1 1 1 | 1 1 1 | 1 1 1 | 1 1 1 | 1 0 | 1 0 |
| 37.5° | 1 1 1 | 1 1 1 | 1 1 1 | 1 1 1 | 1 1 1 | 1 1 1 | 1 0 |
| 52.5° | 0 0 1 | 0 1 | 0 1 | 0 1 | 0 1 | 0 1 | 0 1 |
| 67.5° | 0 0 1 | 0 0 1 | 0 1 | 0 1 | 0 1 | 0 1 | 0 1 |
| 82.5° | 0 0 1 | 0 0 1 | 0 0 1 | 0 1 | 0 1 | 0 1 | 0 1 |
| M totals | 12 | 10 | 8 | 6 | 4 | 2 | 0 |

The reason for this procedure is that the photons appear at random time intervals, and with random values of polarization $\phi$ (unlike the orderly entries of Table 18-1). The easiest way to handle the random stream of data is to use the coincidence counter of Fig. 18-1. The counter gives the number of matches M (0,0 + 1,1) and also the number of mismatches (0,1 + 1,0).

A final note concerning the equipment: the calcite filter's output is useless unless it can be converted into an electrical signal. Accordingly, each filter feeds a detector in the form of a photomultiplier. The latter is sensitive enough to respond to a reasonable fraction of entering photons. In practice, one must use two photomultipliers, one for the 1 output and the other for the 0 output. To simplify the diagram, however, a single "photon detector" block is shown; it merely converts the filter's 0s and 1s into electrical 0s and 1s.

### Experiment Using Calcite Filters

Now consider the gathering of typical experimental data [36, 37]. In the following numerical example, $\theta_B$ is set to 30°. Starting at $t = 0$, because $\phi$ randomly varies between −90° and +90°, we get a string of 0s and 1s. In Fig. 18-1, the A output is 1 0 0...0 0 1. With $\theta = 30°$, the B output is 1 0 1...0 0 1. Out of the string of 16 binary digits, the coincidence counter shows that M (the number of matches) is 12, corresponding to M = 75%.

What do we expect? Since $\phi$ is a random variable between −90° and + 90°, we can take representative samples 15° apart, say, such as depicted in the rows of Table 18-1. The columns represent values for the calcite filter difference angle $\theta$ = 0°, 15°,..., 90°. The A columns represent the output of the A filter, which remains the same for the entire table because it is not rotated. The B columns, however, rotate down one row one row distance, $\phi = 15°$, as we move to the right one column distance, $\theta = 15°$. The M totals appear to follow a linear decrease as $\theta$ linearly increases.

The following examples illustrate some of the Table 18-1 entries in the $\theta = 75°$ columns:

$\phi = -82.5°$, $\theta_A = 0°$,      $|\phi - \theta| = 82.5°$, A output = 0;

$\phi = -82.5°$, $\theta_B = \theta = 75°$,      $|\phi - \theta| = 157.5°$, B output = 1 (M = 0)

$\phi = -22.5°$, $\theta_A = 0°$,      $|\phi - \theta| = 22.5°$, A output = 1;

$\phi = -22.5°$, $\theta_B = \theta = 75°$,      $|\phi - \theta| = 97.5°$, B output = 0 (M = 0)

$\phi = 52.5°$, $\theta_A = 0°$,      $|\phi - \theta| = 52.5°$, A output = 0;

$\phi = 52.5°$, $\theta_B = \theta = 75°$,      $|\phi - \theta| = 22.5°$, B output = 1 (M = 0).

The straight-line plot of $M_{total}$ as a function of $\theta$ is shown as the "expected" curve in Fig. 18-2. The "measured" curve is also shown; it is given by $M_{total} = 12\cos^2\theta$. Quantum theory, as usual, agrees with the experimental observations by predicting a variation that also has a $\cos^2\theta$ shape.

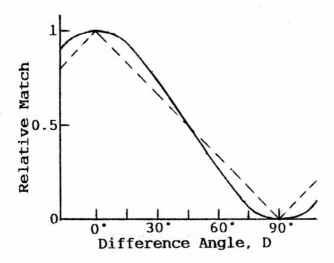

**Fig. 18-2.** For the experiment of Fig. 18-1, this shows the matches versus the calcite filters' difference angle, θ. The dashed curve shows the relative expected matches as given by Table 18-1, while the solid curve shows the relative values actually measured, $\cos^2\theta$.

Bell's assertion that superluminal effects are occurring is based on the curves of Fig. 18-2. At $\theta = 15°$, for example, we expect a coincidence probability of M = 10/12 = 83%; instead, we measure a probability of M = 11.2/12 = 93%. At $\theta = 30°$, we expect M = 8/12 = 67%; instead, we measure M = 9/12 = 75%, and so forth. Since the outputs of calcite filters A and B have a much higher coincidence than expected, even though they are physically very far apart, there must be, *somehow*, instantaneous communication between the calcite filters (so the argument goes). My argument, however, is that this nonsensical conclusion only *proves* that there is something wrong with Table 18-1.

Among people who pay attention to such matters, the discrepancy between measured and expected curves of Fig. 18-2 has become a traumatic experience. Visualize the following scenario: the calcite filters in Fig. 18-1 are $10^6$ meters apart. The experimenter, Z, rotates calcite filter B to the $\theta = 0$ position. The coincidence counter reads "M = 100%." So far so good. Then Z rotates filter B to the $\theta = 15°$ position. Z expects "M = 83%"; instead, Z gets "M = 93%." Z checks everything carefully, but there are no errors. The conclusion is inescapable that, at the 15° setting, the A and B photons are connected to each other through a medium that is $10^6$ meters long. Superluminal effects!

Since photons are minuscule, the connection between A and B must be some sort of "string" or "cable." To avoid offending the cosmological "string theory" people, I will call the inter-photon connection a "cable." This may offend electrical engineers, but they will not take the cable proposal too seriously, and will rapidly "hang up" on it.

The photons somehow communicate with each other superluminally through the cable. When photon A exits calcite filter A along the "output 1" path, it *instantaneously* tells this to photon B; the latter, if it was headed for the "output 0" path of filter B, *instantaneously* "changes its mind" and exits along the "output 1" path. Similarly, when photon A exits filter A along the "output 0" path, it instructs photon B, if it was headed for the "output 1" path, to instantaneously change its mind and exit along the "output 0" path of filter B.

Well, almost. Let us not be unrealistic by expecting perfect agreement between photons A and B. If the discrepancy between their calcite positions is too large, one may exit through "output 1" while the other ends up going through "output 0." However, 93% of the time, when $\theta = 15°$, their calcite exits are in agreement.

What is the reaction of physicists to the news that photons A and B can instantaneously control each other's movements over vast distances? Bell's theorem unleashed a tremendous amount of work: theoretical, experimental, and what one may call philosophical. Experimentally, the discrepancy between measured and expected curves of Fig. 18-2 has been verified beyond doubt. In the next section, however, I present a conjecture for explaining how M = 93%, at $\theta = 15°$, without resorting to superluminal message velocities along semi-infinitely-long cables (and without resorting to extrasensory perception, ectoplasm, and so forth).

## A Conjecture That Can Explain the Discrepancy

The difficulty with Table 18-1 is that its entries demonstrate "all-or-nothing" behavior. Suppose that the experiment is conducted, with $\theta_A = \theta_B = 0°$, until 1000 photon pairs are generated in the twin-state block of Fig. 18-1. The polarization $\phi$ of 500 pairs will fall between $-45°$ and $+45°$, and Table 18-1 would have us believe that *each* of them yields A = B = 1, M = 1; for the other 500 pairs, the polarization magnitude is greater than $45°$, so *each* of them yields A = B = 0, and again M = 1.

On the other hand, consider the rough treatment that an *individual photon* suffers as it travels through a calcite filter: its $\underline{E}$ (and $\underline{H}$) lines are rotated, *by as much as 45°*, until the polarization of the internal ray agrees with that of the filter. What I am leading up to, in other words, is that the filter is somewhat imperfect, and the polarization angle of a photon is not a sacred, inviolate constant. It is a trivial matter, in waveguide structures, to change the polarization angle by as much as we please. Simply take a long section of the waveguide of Fig. 4-1, and *gradually* twist it so that the rotation, per cycle of EMF, is reasonably small. It is quite common, in waveguide assemblies, for one reason or another, to require a polarization angle rotation of 90°.

My conjecture is that, because of the "rough treatment," the photon's polarization angle is subject to small perturbations ($\pm 7.5°$ out of $180°$, or $\pm 4.2\%$). Recall the models of Fig. 12-1 and Fig. 12-2: a photon is preceded by a (nominal) compression shock wave as it plows through the aether at the speed of light, and the aether contains streamlines that can guide the photon, depending on slit openings and interference effects. The conjecture is that, in addition to lateral push in an interference apparatus, the ethereal streamlines can *slightly* rotate the photon's plane of polarization. In Fig. 18-1, this can even happen in a short flight on the way to the calcite filter, and/or it can occur inside the filter. (The latter possibility seems more reasonable to me.) Whether, and by how much, $\phi$ is rotated depends on the statistically random but predetermined history of the photon. (Remember, also, that the twin-state photons have appreciably different frequencies and energies.)

In Fig. 18-1, let us suppose that $\phi_A$ and $\phi_B$ can each change by, say, $7.5°$. What does this do to the "expected" curve of Fig. 18-2? To investigate this in a way that is tractable, suppose that $\phi_A$ and $\phi_B$ each randomly and *independently* switch $\pm 7.5°$ with respect to their nominal angle. The probability is 1/4, then, that each of the following four combinations will occur:

$\phi_A + 7.5°$, $\phi_B + 7.5°$, $\phi$ difference = 0;
$\phi_A + 7.5°$, $\phi_B - 7.5°$, $\phi$ difference = 15°;
$\phi_A - 7.5°$, $\phi_B + 7.5°$, $\phi$ difference = −15°;
$\phi_A - 7.5°$, $\phi_B - 7.5°$, $\phi$ difference = 0.

This is equivalent to switching the calcite filter difference angle, $\theta$, as follows:

25% of the time, add 15° to $\theta$;          }
25% of the time, subtract 15° from $\theta$; } (18-1)
50% of the time, no change in $\theta$.        }

In Fig. 18-2, the above is equivalent to moving the curve to the right by 15° [represented by a dashed line in Fig. 18-3(a)]; to the left by 15° [represented by a dot-dash line in Fig. 18-3(a)]; and leaving it alone [represented by a solid line in Fig. 18-3(a)]. When we add the M values in accordance with Equation 18-1, we get the piecewise-linear curve of Fig. 18-3(b).

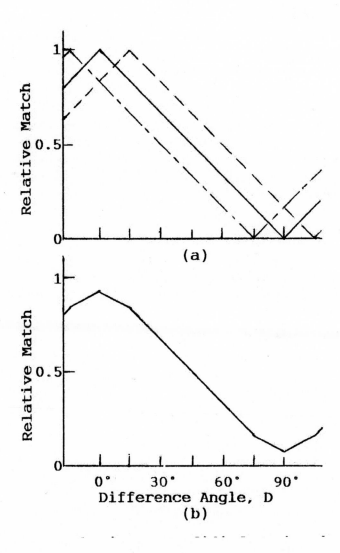

**Fig. 18-3.** Showing how the "expected" curve of Fig. 18-2 is modified if the φ polarization angles of photons A and B randomly shift by ±7.5° before they leave their calcite filters. (a) The shifts shown are equivalent to moving the curve 15° to the right (— — —), or 15° to the left (— · — -), or leaving it alone (———). (b) This shows the result if the ordinate values of (a) are added in accordance with Equation 18-1.

The curve of Fig. 18-3(b) is reasonably similar, in *shape*, to the $\cos^2\theta$ "measured" curve of Fig. 18-2. The main point of the above exercise is to show that small ($\pm7.5°$) random shifts in the polarization angle, if they occur before the twin photons reach their detectors, can approximate the $\cos^2\theta$ function. The experimenter has no way of knowing that the polarization angle of photon A disagrees with that of photon B. This leads to the false and impossible conclusion that the twin photons are instantaneously communicating with each other so as to obtain greater than expected correlation, or probability of matches, when $|\theta| < 45°$.

It seems to me that the notion that one photon can instantaneously influence another is, to repeat, nonsense. Instead of thinking that this is "somehow" possible, we should seek to escape the trap set by Bell's theorem.

## Comment

Finally, the writings of physicist Tore Wessel-Berg are pertinent. An excerpt from *Electromagnetic and Quantum Measurements: A Bitemporal Neoclassical Theory* is reproduced here with the kind permission of Springer Science and Business Media [26] (pp. 327–328):

> It is well documented that Albert Einstein never accepted quantum mechanics as a *complete* description of nature, and he and Niels Bohr debated the quantum reality question throughout their lives. In the famous publication [A. Einstein, B. Podolsky, and N. Rosen, "Can a Quantum-Mechanical Description of Physical Reality be Considered Complete?" *Physical Review* 47 (May 1935): 777–780.] Einstein and his coauthors presented the *EPR* paradox in an attempt to show the incompleteness of quantum theory. They presented a "Gedanken" experiment involving two momentum-correlated electrons, which is conceptually equivalent to the polarization-correlated photon experiment...The idea was to demonstrate that some kind of "hidden variable" had to be added in to explain the results.
>
> Quantum mechanics, represented by Bohr, held the view that the photon polarization does not exist *before* it is measured. Polarization is a relational attribute that does not come into existence until some apparatus is deployed to measure it. The probabilistic polarization wave function *collapses* to either one of the two possible values permitted by the polarizer, vertical or horizontal. Einstein held the realistic view, in line with his profound notion of nature, that the photons have a *definite*, but unknown polarization prior to measurement, reflecting a *classical* ignorance rather than a *quantum* ignorance. His argument on hidden vari-

ables goes as follows. It is an experimental fact that the two photons, with specified polarization $\phi$ and $-\phi$ with regard to the presumably aligned polarizers trigger either both the vertical or both the horizontal counters. If the quantum concept of complete nonexistence of photon polarization before measurement were correct, the vertical or horizontal triggering of the two would be completely *uncorrelated*. But this is contrary to experiments that show that they are indeed completely *correlated*. And then comes Einstein's crucial argument. If we assume that there is no information flowing between the two polarizers, and this information would certainly have to be *instantaneous*, the correlations can only be explained by a *polarization attribute*, implying a functional relationship between the photon polarization $\phi$ and the selection by the photon of vertical or horizontal counters. This "hidden variable" is required to explain the experimental results of complete coincidence between right and left counters. This, in brief, is the *EPR* paradox.

Bohr explains the problem away fairly vaguely in terms of correlation entanglement between the photon pairs, involving in some ways the experimental arrangement. He did not offer a convincing solution to the *EPR* paradox, which has remained an ongoing debate until this day. Einstein's *EPR* argument did not convince supporters of the orthodox quantum interpretation, but this seems to be due more to the obscurity of Bohr's response rather than to any good arguments it brought forward. The reader interested in the detailed history of this decisive point in physics should refer to the book by Mara Beller [Mara Beller, *Quantum Dialogue: The Making of a Revolution* (Chicago: University of Chicago Press, 1999)], which presents an exceedingly well documented and comprehensive discussion of the circumstances surrounding the dialogue between the two, as well as the subsequent development of the Copenhagen Interpretation into its widespread acceptance by a large fraction of the physics community. The account leaves the definite impression that its promotion was more due to the persistent and at times aggressive campaigning persuaded by Bohr and followers, rather than its substance.

My comment is that the "hidden variable" is, of course, *the aether,* which was, ironically, abandoned by Albert Einstein [38, 39]. However, still another explanation, via bitemporal neoclassical theory, is presented in Professor Wessel-Berg's book.

# Appendix

▼

# Some Numerical Values

By definition, a photon always travels at the speed of light (symbol $c$) in a vacuum. Table A-1 illustrates various frequencies (and corresponding wavelengths), ranging from those of power stations (60 Hz) to gamma rays ($3 \times 10^{21}$ Hz). The wavelength entries are correct only for a vacuum; in any other medium, the velocity of propagation is less than $c$ and depends on the medium.

**Table A-1.** Various electromagnetic frequencies and corresponding wavelengths in vacuum. (Frequency × Wavelength = $c \cong 3 \times 10^8$ m/s.)

| Frequency | Wavelength | Application |
|---|---|---|
| 60 Hz | 5000 kilometers | Power stations in United States |
| 3 kHz = $3 \times 10^3$ Hz | 100 kilometers | Approx. low-freq. edge of EMF signaling |
| 1 MHz = $1 \times 10^6$ Hz | 300 meters | Approx. center of AM broadcast band |
| 100 MHz = $1 \times 10^8$ Hz | 3 meters | Approx. center of FM broadcast band |
| 300 GHz = $3 \times 10^{11}$ Hz | 1 millimeter | Approx. low-freq. edge of infrared |
| $4 \times 10^{14}$ Hz | 750 nanometers | Low-frequency edge of visible light |
| $7.9 \times 10^{14}$ Hz | 380 nanometers | High-frequency edge of visible light |
| $3 \times 10^{17}$ Hz | 1 nanometer | Approx. borderline of ultraviolet, X-rays |
| $6 \times 10^{18}$ Hz | 0.05 nanometer | Typical X-ray |
| $1 \times 10^{20}$ Hz | 3 picometers | Approx. borderline of X-rays, gamma rays |
| $3 \times 10^{21}$ Hz | 0.1 picometer | Typical gamma ray |

Some of the numerical values used (or, in some cases, derived) in this book, along with symbols, where appropriate, are given in Table A-2. Most of the values are taken from a small booklet, *Particle Physics*, that is published by the American Institute of Physics [40].

**Table A-2.** Various numerical values.

| Name | Symbol | Name | Symbol |
|------|--------|------|--------|
| Meter | m | Coulomb | C |
| Second | s | Farad | F |
| Joule | J | Henry | H |
| Newton | N | Ohm | $\Omega$ |
| Kilogram | kg | Year | yr |
| Kelvin | K | Light-year | lt-yr |

| Name | Symbol | Numerical value |
|------|--------|-----------------|
| Speed of light in vacuum | $c$ | $2.9979 \times 10^8$ m/s |
| Planck's constant | $h$ | $6.6261 \times 10^{-34}$ J·s |
| Gravitational constant | $G$ | $6.6726 \times 10^{-11}$ N·m$^2$/kg$^2$ |
| Boltzmann's constant | $k_B$ | $1.38066 \times 10^{-23}$ J/K |
| Permittivity of vacuum | $\varepsilon_0$ | $8.8542 \times 10^{-12}$ F/m |
| Electrostatic constant | $k$ | $8.9876 \times 10^9$ N·m$^2$/C$^2$ |
| Permeability of vacuum | $\mu_0$ | $4\pi \times 10^{-7}$ H/m |
| Electron's charge | $e$ | $1.60218 \times 10^{-19}$ C |
| Electron's mass | $m_0$ | $9.1094 \times 10^{-31}$ kg |
| Proton's mass | | $1.67262 \times 10^{-27}$ kg |
| Neutron's mass | | $1.67493 \times 10^{-27}$ kg |
| Sun's mass | $M$ | $1.988 \times 10^{30}$ kg |
| Earth's mass | | $5.974 \times 10^{24}$ kg |
| Mean radius of Earth's orbit | | $1.496 \times 10^{11}$ m |
| Characteristic imp. of vacuum | $Z_0$ | $376.7$ $\Omega$ |
| Range of strong force | | $1.4 \times 10^{-15}$ m |
| Year | yr | $3.1558 \times 10^7$ s |
| Light-year | lt-yr | $9.4605 \times 10^{15}$ m |
| Parsec | | $3.262$ lt-yr |

The three EMF mediums listed in Table 3-1 are nonmagnetic, so each has the same permeability, $\mu = 4\pi \times 10^{-7} = 1.257 \times 10^{-6}$ henries/meter. The reciprocal of permittivity, $1/\varepsilon$, is listed. For vacuum or air $1/\varepsilon = 1/8.8542 \times 10^{-12} = 1.129 \times 10^{11}$ meters/farad. (For ruby mica and water, the $1/\varepsilon$ values are divided by their dielectric constants, 5.4 and 78 respectively.)

One of the strongest arguments for an aether is that the so-called vacuum has measurable characteristics, such as velocity of propagation and characteristic impedance and, above all, that sound has analogous properties. Let's consider the two right-hand columns in Table 3-1: the derived values for $v$ and $Z_0$. For sound, velocity is given by

$$v = (Y_0/\rho_0)^{1/2}, \quad \text{(A-1)}$$

while for an EMF we have

$$v = 1/(\mu\varepsilon)^{1/2}. \quad \text{(A-2)}$$

These equations were used to derive the listed values. Notice that, for an EMF propagating in a vacuum, we get $v = 2.998 \times 10^8$ m/s.

For sound, the characteristic impedance is given by

$$Z_0 = (\rho_D Y_0)^{1/2}, \quad \text{(A-3)}$$

while for an EMF we have

$$Z_0 = (\mu/\varepsilon)^{1/2}. \quad \text{(A-4)}$$

Notice that, for an EMF in vacuum, the well-known value 376.7 ohms is obtained. Since $Z_0$ for ruby mica and water is different from that of vacuum or air, an EMF traveling from air to mica, or from air to water, is partially reflected (and partially transmitted) at the boundary between the dissimilar mediums.

As the universe expands, the density of aether particles may decrease, accompanied by changes in permeability and/or permittivity. The corresponding changes in the velocity of light imply that $c$ may have been different in the past, and may be different in the future (along with many of the other natural "constants"). This is an unresolved nightmare for astrophysicists.

# REFERENCES

[1] Whittaker, E. T. *A History of the Theories of Aether and Electricity*. New York: Thomas Nelson and Sons, 1951.

[2] Miller, D. C. "The Aether-Drift Experiment and the Determination of the Absolute Motion of the Earth." *Rev Mod Phys* 5 (July 1933): 203–242.

[3] Taseja, T. S., A. Javan, J. Murray, and C. H. Townes. "Test of Special Relativity or of the Isotropy of Space by Use of Infrared Masers." *Phys Rev* 133 (March 1964): A1221–A1225.

[4] Brillet, A., and J. L. Hall. "Improved Laser Test of the Isotropy of Space." *Phys Rev Lett* 42 (February 1979): 549–552.

[5] Drever, R. W. P., J. L. Hall, F. V. Kowalski, J. Hough, G. M. Ford, A. J. Munley, and H. Ward. "Laser Phase and Frequency Stabilization Using an Optical Resonator." *Appl Phys B* 31 (1983): 97–105.

[6] Wolf, P., S. Bize, A. Clairon, A. N. Luiten, G. Santarelli, and M. E. Tobar. "Tests of Lorentz Invariance Using a Microwave Resonator." *Phys Rev Lett* 90 (Feb 2003): 060402-1-4.

[7] Muller, H., S. Herrmann, C. Braxmaier, S. Schiller, and A. Peters. "Modern Michelson-Morley Experiment Using Cryogenic Optical Resonators." *Phys Rev Lett* 91 (July 2003): 020401-1-4.

[8] Consoli, M., and E. Costanzo. "From Classical to Modern Aether-Drift Experiments: The Narrow Window for a Preferred Frame." *Phys Lett A* 333 (2004): 355–363.

[9] Antonini, P., M. Okhapkin, E. Goklu, and S. Schiller. "Test of Constancy of Speed of Light With Rotating Cryogenic Optical Resonators." *Phys Rev A* 71 (2005): 050101.

[10] Lineweaver, C. H., and T. M. Davis. "Misconceptions About the Big Bang." *Sci Am* 292 (March 2005): 36–45.

[11] Mermin, N. D. *Space and Time in Special Relativity*. New York: McGraw-Hill, 1968.

[12] Mills, R. *Space, Time and Quanta*. New York: Freeman, 1994.

[13] Admirers of Maurice Allais. "About the Aether Concept." *Maurice Allais: The Scientist*. http://allais.maurice.free.fr/English/aether1.htm.

[14] Krauss, L. M., and M. S. Turner. "A Cosmic Conundrum." *Sci Am* 291 (September 2004): 71–77.

[15] Galison, P. *Einstein's Clocks, Poincaré's Maps*. New York: Norton, 2003.

[16] Einstein, A. "On the Electrodynamics of Moving Bodies." *Annalen der Physik* 17 (1905): 891–921.

[17] Blanchard, C. H., C. R. Burnett, R. G. Stoner, and R. L. Weber. *Introduction to Modern Physics*. 2nd ed. Englewood Cliffs, NJ: Prentice-Hall, 1969.

[18] Einstein, A. "Aether and the Theory of Relativity." *Theories of the Aether*. http://www.mountainman.com.au/aether_0.html. There are typographical errors. The most serious of which is that paragraph 16 should say "Generalizing, we must say this:…but the hypothesis of aether in itself is not in conflict with the special theory of relativity."

[19] Panofsky, W. K. H., and M. Phillips. *Classical Electricity and Magnetism*. Reading, MA: Addison-Wesley, 1955.

[20] Milton, R. "Aether Drift?" Alternative Science Web site, 1994 (inactive).

[21] Freeman, K. C. "The Hunt for Dark Matter in Galaxies." *Science* 302 (December 2003): 1902.

[22] Deutsch, S. *Return of the Aether*. Mendham, NJ: SciTech, 1999.

[23] Magueijo, J. *Faster Than the Speed of Light*. Cambridge, MA: Perseus, 2003.

[24] Tonomura, A., J. Endo, T. Matsuda, T. Kawasaki, and H. Ezawa. "Demonstration of Single-Electron Buildup of an Interference Pattern." *Am J Phys* 57 (February 1989): 117–120.

[25] Smythe, W. R. "Aether Hypothesis." In *McGraw-Hill Encyclopedia of Physics*, edited by S. P. Parker, 392. 2nd ed. New York: McGraw-Hill, 1993.

[26] Wessel-Berg, T. *Electromagnetic and Quantum Measurements: A Bitemporal Neoclassical Theory*. Boston: Kluwer, 2001.

[27] Herbert, N. *Quantum Reality.* New York: Anchor, 1985.

[28] Baggott, J. *The Meaning of Quantum Theory.* Oxford: Oxford University Press, 1992.

[29] Lindley, D. *Where Does the Weirdness Go?* New York: Basic Books, 1996.

[30] Lamoreaux, S. K. "Demonstration of the Casimir Force in the 0.6 to 6 µm Range." *Physical Review Letters* 78 (January 1997): 5–8.

[31] Bohm, D., and B. Hiley. *The Undivided Universe: An Ontological Interpretation of Quantum Mechanics.* London: Routledge, 1993.

[32] Kwiat, P., H. Weinfurter, and A. Zeilinger. "Quantum Seeing in the Dark." *Sci Am* 275 (November 1996): 72–78.

[33] Albert, D. Z. "Bohm's Alternative to Quantum Mechanics." *Sci Am* 270 (May 1994): 58–67.

[34] Bell, J. S. "On the Einstein Podolsky Rosen Paradox." *Physics* 1 (1964): 195–200.

[35] Bell, J. S. *Speakable and Unspeakable in Quantum Mechanics.* Cambridge: Cambridge University Press, 1987.

[36] Clauser, J. F., and A. Shimony. "Bell's Theorem: Experimental Tests and Implications." *Reports on Progress in Physics* 41 (1978): 1881–1927.

[37] Aspect, A., J. Dalibard, and G. Roger. "Experimental Test of Bell's Inequalities Using Time-Varying Analyzers." *Physical Review Letters* 49 (1982): 1804–1807.

[38] Rarity, J. G. "Getting Entangled in Free Space." *Science* 301 (August 2003): 604–605.

[39] Giovannetti, V., S. Lloyd, and L. Maccone. "Quantum-Enhanced Measurements: Beating the Standard Quantum Limit." *Science* 306 (November 2004): 1330–1336.

[40] *Particle Physics.* American Institute of Physics, July 1994.

[41] Chown, M. "Catching the Cosmic Wind." *New Scientist* 186 (April 2005): 30–33.

# BIOGRAPHICAL SKETCH

Sid Deutsch received a BEE degree in 1941 from Cooper Union and a PhD in 1955 from what is now Polytechnic University. He taught electrical engineering courses at the following institutions:

Polytechnic University (1955–1972);

Rutgers University (1972–1979);

Tel-Aviv University (1979–1983);

University of South Florida (1983–1998).

He is a Fellow of the IEEE and the Society for Information Display.

He has also written or co-authored seven books:

Deutsch, S. *Theory and Design of Television Receivers*. New York: McGraw-Hill, 1951.

Deutsch, S. *Models of the Nervous System*. New York: John Wiley, 1967.

Welkowitz, W., and S. Deutsch. *Biomedical Instruments: Theory and Design*. New York: Academic Press, 1976.

Deutsch, S., and E. Tzanakou. *Neuroelectric Systems*. New York: New York University Press, 1987.

Deutsch, S., and A. Deutsch. *Understanding the Nervous System: An Engineering Perspective*. New York: IEEE Press, 1993.

Deutsch, S. *Return of the Ether*. Mendham, NJ: SciTech Publishing, 1999.

Deutsch, S. *Are You Conscious, and Can You Prove It?* iUniverse, 2003.

# INDEX

**A**

Aether,
    density, 2, 31
    drift, 2, 4, 6, 7, 11, 30, 35
    elasticity, 24
    particle, 32, 69, 72
    shock wave, 87
    spin, 32, 69
    wind, 74, 115
Air, 21
Albert, D. Z., 121, 149
Allais, M., 10, 148
Antonini, P., 42, 147

**B**

Bell's theorem, 125
Birefringent, 128
Blue shift, 3, 6
Bohr, N., 138
Brillet, A., 38, 147

**C**

Calcite filter, 127, 129
Capacitance, 28
Casimir effect, 92
Cathode-ray tube, 104
Cavity resonator, 38, 42
Characteristic impedance, 21, 145

978-0-595-37481-6
0-595-37481-6

Printed in the United States
45242LVS00006B/53